DISTRIBUTED SYSTEMS SECURITY

DISTRIBUTED SYSTEMS SECURITY
Issues, Processes and Solutions

Abhijit Belapurkar, *Yahoo! Software Development India Pvt. Ltd., India*
Anirban Chakrabarti, *Infosys Technologies Ltd., India*
Harigopal Ponnapalli, *Infosys Technologies Ltd., India*
Niranjan Varadarajan, *Infosys Technologies Ltd., India*
Srinivas Padmanabhuni, *Infosys Technologies Ltd., India*
Srikanth Sundarrajan, *Infosys Technologies Ltd., India*

A John Wiley and Sons, Ltd., Publication

This edition first published 2009
© 2009 John Wiley & Sons Ltd

Registered office
John Wiley & Sons Ltd, The Atrium, Southern Gate, Chichester, West Sussex, PO19 8SQ, United Kingdom

For details of our global editorial offices, for customer services and for information about how to apply for permission to reuse the copyright material in this book please see our website at www.wiley.com.

Other Wiley Editorial Offices

John Wiley & Sons Inc., 111 River Street, Hoboken, NJ 07030, USA
Jossey-Bass, 989 Market Street, San Francisco, CA 94103-1741, USA
Wiley-VCH Verlag GmbH, Boschstr. 12, D-69469 Weinheim, Germany
John Wiley & Sons Australia Ltd, 42 McDougall Street, Milton, Queensland 4064, Australia
John Wiley & Sons (Asia) Pte Ltd, 2 Clementi Loop #02-01, Jin Xing Distripark, Singapore 129809
John Wiley & Sons Canada Ltd, 6045 Freemont Blvd, Mississauga, ONT, L5R 4J3, Canada

Wiley also publishes its books in a variety of electronic formats. Some content that appears in print may not be available in electronic books.

The authorship of the book by Abhijit Belapurkar is in no way by, for, or in the name of Yahoo! and the views expressed in the book are exclusively those of Abhijit and other coauthors and not Yahoo's.

Library of Congress Cataloging-in-Publication Data

Distributed systems security issues, processes, and solutions / Abhijit Belapurkar ... [*et al.*].
    p. cm.
  Includes bibliographical references and index.
  ISBN 978-0-470-51988-2 (cloth)
  1. Computer Security. 2. Electronic data processing – Distributed processing. 3. Internet – Security measures. I. Belapurkar, Abhijit.
    QA76.9.A25D567 2009
    005.8 – dc22

                                                                                    2008034570

A catalogue record for this book is available from the British Library.

ISBN 978-0-470-51988-2   (H/B)

Typeset in 11/13 Times by Laserwords Private Limited, Chennai, India
Printed in Great Britain by CPI Antony Rowe, Chippenham, Wiltshire

In memory of

Late Dr. Anirban Chakrabarti

Our esteemed colleague and co-author Anirban Chakrabarti., Ph.D passed away on 7th September 2008 in an accident leaving a void in his family, friends as well as colleagues. We deeply mourn Anirban's untimely death and pray for his soul. Anirban is survived by his wife Lopa, a 11 month old son Ishaan and mother. We extend our deepest condolences as well as support to the mourning family.

Anirban was a Principal Researcher and Head of the Grid Computing Research Group in Software Engineering Technology Labs (SETLabs) of Infosys Technologies, India. Anirban holds a Bachelor's in Engineering degree from Jadavpur University, India, and a Ph.D. degree from Iowa State University, USA. Anirban has been an active researcher in conferences like HiPC, ADCOM, and INFOCOM. Prior to this book he authored a book titled *"Grid Computing Security"* in 2006 (published by Springer). Anirban received the "Research Excellence Award" from Iowa State University in 2003 and the Infosys Excellence Awards in 2006 and 2008.

Contents

List of Figures

List of Tables

Foreword

The area of information security is a classic example of a human endeavour where the theorists and practitioners are completely polarized. This emanates from the myth that cryptography and information security are one and the same. While cryptography is an essential component of information security, it is not an end in itself. The encryption of a message may ensure its secure passage through a network but not at the end-points. The advent of Internet resulted in the development of the secured socket layer protocol that only catered to the movement of hypertext securely over a public network.

Around the turn of the new millennium, a new disruptive technology called the Web Services emerged. It was a simple and beautiful idea: aligning self-contained business functionalities in the form of software components that could be published, found and consumed programmatically. On the technical front, interoperability became the buzzword; XML became the lingua franca for silicon-based life forms. The published interfaces replaced the APIs. Web Services were followed by the generic Service-Oriented Architecture. This called for a paradigm shift in thinking about architecture, software transactions and information security. Taking a cue from the information security text books, it no longer remained a Bob and Alice issue – it became a Bob, Alice, Ted, Carroll and others issue.

Contemporaneous to the development of SOA, the rise of high-performance or grid computing is another important milestone. The grid consists of loosely-coupled systems that work in unison to carry out computationally-intensive tasks. It also employs the principle of CPU scavenging. One serious security challenge is due to the presence of untrustworthy systems acting as malicious nodes.

This book covers the entire secure software development lifecycle process – from requirements analysis to testing and implementation. In addition, it also looks at the abstract picture from an Enterprise IT point of view. It follows a layered approach: hosts, infrastructure, applications and services. The vulnerabilities and threats as well as the solutions for each layer form the backbone of this book. For the sake of completeness, the authors have made a serious attempt to discuss the four basic pillars of information security in terms of issues and techniques keeping in mind the typical software developer. The real highlight of the book is the inclusion of security standards for distributed systems that have been developed over the last eight years. The book includes a compliance case study involving policies and identity management as well as a case study concerning the grid. Finally, the authors provide us a sneak preview into the future through the coverage of security issues around Cloud Computing, the emerging area of Usercentric Identity Management and a relatively new cryptosystem called the Identity-Based Encryption.

I firmly believe that this book is a treasure for those practitioners who are involved in design, implementation and deployment of secured distributed systems.

Hemant Adarkar, PhD
Enterprise Architect

Preface

Overview

As we move more and more to a better-connected world, systems are becoming more distributed in terms of geography as well as functionality. The phenomenon of distributed systems and computing is becoming increasingly relevant in a consumer world in which social networking sites like Orkut, Facebook and so on are becoming tremendously popular, with the user count crossing tens of millions in a few years of their existence. Enterprises are now witnessing increasing collaboration and data sharing among the different participating entities, resulting in the need for and use of distributed resources and computing. Another important element that has increased the complexity of IT operations is the need for integration of different applications: middleware developed in different platforms and by different vendors. We are also seeing a spurt of mergers and acquisitions which require integration of technologies across enterprises. Moreover, the enterprises are outsourcing the nonessential elements of the IT infrastructure to various forms of service provider. Distributed computing is therefore a necessity that most enterprises are embracing.

Distributed computing technologies followed a very classical pattern of evolution. They were initiated in the academic and research communities, to fulfill the need to connect and collaborate, and slowly they were adopted by the enterprises. Presently, enterprises and user communities cannot live without some application of distributed computing. However, with the widespread adoption of distributed computing, experts are pointing out the security issues that can hurt these enterprises and user communities in a huge way. Analyzing the security issues and solutions in distributed computing is not simple. Different solutions exist and hence it is necessary to identify the different layers of the distributed computing environment and analyze the security issues in a holistic manner. In this book, *Distributed Systems Security*, we provide a holistic insight into current security issues, processes and solutions, and map out future directions in the context of today's distributed systems. This insight is elucidated by modeling of modern-day distributed systems using a four-tier logical model: host layer, infrastructure layer, application layer and service layer (bottom to top). We provide an in-depth coverage of security threats and issues across these tiers. Additionally, we describe the approaches required for efficient security engineering, as well as exploring how existing solutions can be leveraged or enhanced to proactively meet the dynamic needs of security for the next-generation distributed systems. The practical issues thereof are reinforced via practical case studies.

Organization

In this book we have made very few assumptions on the prerequisites for readers. In the different sections, we have provided sufficient information and background material for readers new to this area. The book is organized into fourteen chapters. In Chapter 1, we provide a brief overview of distributed systems. We felt the need to inform readers about the general issues in distributed systems, before delving deep into the security aspects. We talk about the characteristics and different types of distributed system, and also provide an overview of challenges faced in this area. Though challenges like synchronization and fault tolerance are critical, due to the explosive growth of distributed systems and their complexities, the security challenge is paramount. In this chapter, we also provide a brief motivation for the layered approach to dissecting distributed systems. Finally, we provide a list of trends in distributed systems security.

In Chapter 2 we talk about the diverse security engineering aspects. We stress that security is to be treated as an integral part of the software development lifecycle (SDLC). We provide an overview of some of the prevailing security-aware software development lifecycle process models and processes, including SSE-CMM, Microsoft SDL and CLASP. In terms of the SDLC activities, we cover in detail related security engineering activities including security requirements activities, threat modeling, security architecture and design reviews, code reviews and security testing.

In Chapter 3 we provide an overview of the common security issues and technologies that are relevant to distributed systems. In the first half, we elucidate the typical security concerns of confidentiality, integrity, access control and availability. Additionally, the issues of trust and privacy are explained. In particular, the emerging need for identity management is explored. In the second half, we explore the different technologies typically used to address these security issues, including encryption mechanisms, PKI, firewalls and digital signatures.

From Chapter 4 to Chapter 7, we delve into the threats and vulnerabilities of different layers defined in Chapter 1.

In Chapter 4, look at security threats and vulnerabilities at the host layer. We broadly group the host-level threats into two categories: transient code threats and resident code threats. In the category of transient code vulnerabilities, we cover various malwares including Trojan horses, spyware, worms and viruses. Additionally, under transient code vulnerabilities, we cover threats in the form of eavesdropping, job faults and resource starvation. In the category of resident attacks, we primarily look at overflow attacks, privilege-escalation attacks and injection attacks.

In Chapter 5, we carry the same thread forward by providing details about threats and vulnerabilities in the infrastructure layer. We divide the infrastructure threats and vulnerabilities into three main categories: network threats and vulnerabilities, grid and cluster threats and vulnerabilities, and data systems threats and vulnerabilities. In the first category we talk about denial-of-service (DoS) attacks, domain name server (DNS) attacks, routing attacks, high-speed network threats and wireless threats. In the second category we talk about threats and issues in grid and cluster architecture, infrastructure and management, and also trust. In data systems, we talk about storage area networks (SAN) and distributed file systems (DFS) threats.

In Chapter 6, we talk about application threats and vulnerabilities. We cover in detail the various injection attacks, including SQL injection, LDAP injection and XPath injection attacks. We go on to cover in detail cross-site scripting attacks. We study attacks caused by improper session management or improper error handling, or due to improper use of cryptography. We also describe other attacks, including DOS attacks and attacks caused by insecure configuration, or canonical representation flaws, or buffer overflows.

In Chapter 7, we talk about the diverse service-level issues, threats and vulnerabilities. Key requirements for service-level security include the need to leverage typical mechanisms of encryption and digital signatures while making sure partial-content encryption and signing is possible. Likewise, it is important to note that mechanisms for interoperation of diverse security solutions are essential, as services operate across heterogeneous systems. Hence the need for a standards-based approach to security is highlighted. In the latter half of the chapter, a detailed analysis of the various threats is provided in the context of services. The plaintext nature of XML, the lingua franca of service-based applications, makes attacks on services easier. The majority of these attacks are morphed forms of conventional attacks for services. We provide a detailed classification of the relevant service-level threats in a logical hierarchy, ranging from attacks purely on services, through attacks on the inter-service communication, to service-authentication attacks.

From Chapter 8 to Chapter 11, we talk about different solutions pertaining to the threats and vulnerabilities mentioned before.

In Chapter 8, we look at some of the host-level security solutions relating to isolation, resource management and host protection. The key solutions studied in depth include sandboxing, virtualization, efficient resource management, anti-malware and memory firewalls. In the context of sandboxing, kernel-loadable modules, user-level sandboxing, delegated architectures and file-system isolations are studied. The diverse models of virtualization, including full-system virtual-ization, para virtualization, shared-kernel virtualization and hosted virtualization are studied, and the inherent security offered via isolation is explained. In the context of resource management, techniques like advance reservation and priority reduction are studied. In antimalware, both signature-based scanning and real-time scanning techniques are explored.

In Chapter 9, we talk about solutions in the infrastructure layer. We refer back to the threats categories, namely network, grid and cluster, and data systems. As part of the network solutions, we discuss information security solutions such as Secure Socket Layer (SSL), IP Security (IPSec) and Virtual Private Networks (VPN). We also talk about DoS solutions and research by looking at application filtering, packet filtering, location hiding, logging and other solutions. As part of the DNS solution, we briefly talk about the DNSSec solution. Routing and wireless solutions are dealt with in detail by talking about several existing techniques. As part of the solution to grid security issues, architectural solutions like Grid Security Infrastructure (GSI) are discussed in detail. We also discuss authorization solutions like VO-level authorization systems (e.g. CAS) and resource-level authorization systems (e.g. PERMIS). In addition to these, we discuss management solutions, such as credential-management systems like MyProxy and trust-management sys-tems like TrustBuilder. As part of the security solution for data systems, we talk about Fiber Channel Security Protocol (FC-SP), DFS Security and security in highly-distributed data systems like OceanStore.

In Chapter 10, we talk about industry best practices to help prevent the common application security vulnerabilities discussed in Chapter 6. First, the role of input-validation techniques is explored in depth. Next, secure session management-related best practices are outlined. Also outlined are best practices for cryptography and encryption. Finally, best practices in error handling and input/output filtering for XSS attack prevention are given.

In Chapter 11, we concentrate on different solutions to the diverse service-level issues, and mechanisms to handle these threats and vulnerabilities. First, we explore why SSL, the predominant solution for Web-based systems, is not enough for Web services-based systems. Further, we highlight the role of standards in pro-moting interoperability, a key requirement for service-oriented IT systems. We explore in detail the complete services security standards stack, right from the bottom layers of XML Encryption/Signature to the Federated identity standards. Finally, the emergence of a new breed of firewalls, XML firewalls, is explained, looking at their critical role in addressing various service-level threats. We provide an exhaustive drill-down view of a typical XML firewall, including an outline of the different configurable parameters. We also explore the role of policy-centered security architectures in satisfying key service-oriented security requirements. We then provide a detailed threat-by-threat solution mapping for better elucidation.

One of the key contributions of this book is to come up with a couple of detailed case studies, which we describe in Chapters 12 and 13. In Chapter 12 we talk about a compliance case study in the financial industry. We highlight how a multilevel, policy-based, service-oriented security architecture is suited to solve such a scenario. In Chapter 13 we give a grid case study, where we look again at a financial organization, running its financial applications in a grid environment.

Finally, in Chapter 14, we look into the crystal ball and predict some important security technologies which may assume importance in the future. In this chapter, we talk about cloud computing security, security appliances, usercentric identity management and identity-based encryption (IBE).

Acknowledgments

We would like to thank all the people who have contributed directly and indirectly to the book's development. Special thanks should go to the reviewers, Vishal Dwivedi, Bijoy Majumdar, Anish Damodaran, and several others whose comments have been invaluable in the progress of the book. Moreover, we would like to thank Birgit Gruber, Sarah Hinton, Sarah Tilley and Emily Dungey of Wiley for their help throughout the book-creation process. Finally, we would like to thank our respective families, without whose support the book could not have been completed.

1

Introduction

1.1 Background

In the 1960s, the great science-fiction writer Isaac Asimov [1] predicted a future full of robots, protecting and sometimes controlling human destiny. Fifty years later, a human-like and all-purpose robot still remains a dream of the robotics research community. However, technological progress in the last couple of decades have ensured that human lifestyle, human interactions and collaboration patterns have changed so dramatically that if anyone like Asimov had written about today's world 50 years back, it would have seemed like science fiction. If we compare the interaction and collaboration patterns of today with those of a decade back, we will find stark differences between the two. E-mails, blogs, messengers and so on are common tools used nowadays which were unknown ten years ago. People seldom stand in a queue in a bank; automated teller machines (ATMs) have become an essential commodity. Similarly, credit cards have taken over from cash and cheques as the new mode of transaction. Internets have become the de facto source of information for millions of people. The new technologies have redefined the ways in which interaction and collaboration between different individuals take place, which in turn are creating a new social-interaction methodology. For example, English is fast becoming a lingua franca for the technical community across the world and the interactions of that community are redefining the English language in a significant way. In addition, geographical and cultural borders are slowly disappearing as social networking sites like Orkut [2], Facebook [3] and so on change the ways people interact. Similar changes are also taking place in the enterprise-computing scenario. Until recently, application developers could safely assume that the target environment was homogeneous, secure, reliable and centrally-managed. However, with the advent of different collaborative and data-sharing technologies, new modes of interaction are evolving. These evolutionary pressures generate new requirements

Distributed Systems Security   A. Belapurkar, A. Chakrabarti, S. Padmanabhuni, H. Ponnapalli, N. Varadarajan and S. Sundarrajan © 2009 John Wiley & Sons, Ltd

for distributed application development and deployment. Enterprises are now witnessing increasing collaboration and data sharing among the different participating entities, resulting in the need for and use of distributed resources and computing. Another important element that has increased the complexity of IT operations is the need for integration of different applications, with middleware developed in different platforms and by different vendors. We are also seeing a spurt of mergers and acquisitions which require integration of technologies across enterprises. Moreover, the enterprises are outsourcing the nonessential elements of the IT infrastructure to various forms of service provider. The technologies that have transformed the world so significantly fall under the bracket of *distributed computing* technologies.

Distributed computing technologies follow a similar pattern of interaction, where disparate and sometimes heterogeneous systems interact with one another over a common communication platform. Initiated by the academic and research community to fulfill the need to connect and collaborate, slowly this technology was adopted by enterprises. Finally, enterprises and user communities cannot live without some application of distributed computing. However, with the widespread adoption of distributed computing, experts are pointing out security issues that can hurt the enterprises and user communities in a huge way. Analyzing the security issues and solutions in distributed computing is not simple as there is a need to identify the interactions between different layers of the distributed computing environment. Different solutions exist and it is necessary to identify the different layers of the distributed computing environment and analyze the security issues in a holistic manner. This book is an effort in that direction.

1.2   Distributed Systems

Distributed systems involve the interaction between disparate independent entities, bounded by common language and protocols and working toward a common goal. Different types of distributed systems are found in real life. One of the biggest and perhaps the most complex distributed system is human society itself. In the digital world, the Internet has become a very important distributed environment for everybody.

1.2.1   Characteristics of Distributed Systems

If we look at any distributed system, for example the Internet, there are several mandatory characteristics, in addition to 'good-to-have' or desirable characteristics. Mandatory characteristics determine the basic nature of distributed systems, such as having multiple entities, heterogeneity, concurrency and resource sharing.

(1) *Multiple entities:* One of the key characteristics of a distributed system is the presence of multiple – in many cases a great many – entities participating

in the system. The entities can be users or subsystems which compose the distributed system.

(2) *Heterogeneity:* Another key characteristic is the heterogeneous nature of the entities involved. The heterogeneity may lie in the type of system or user, underlying policies and/or the data/resources that the underlying subsystems consume. The heterogeneity of distributed systems can be best observed in the Internet, where multitudes of systems, protocols, policies and environments interact to create a scalable infrastructure.

(3) *Concurrency:* Another important characteristic that distinguishes any distributed system from a centralized one is concurrency. Different components of distributed systems may run concurrently as the components may be loosely coupled. Therefore there is a need to understand the synchronization issues during the design of distributed systems.

(4) *Resource sharing:* Sharing of resources is another key characteristic of distributed systems.

In addition to the above mandatory characteristics, there are several desirable characteristics for a distributed system.

(1) *Openness:* A desirable characteristic for a distributed system is openness of the underlying architecture, protocols, resources and infrastructure, where they can be extended or replaced without affecting the system behavior. If we look at the Internet, this issue is nicely handled through the use of open standards: we can see the interplay between different protocols, standards, infrastructures and architectures without affecting the activities of the Internet as a whole.

(2) *Scalability:* One of the key motivations for going from a centralized system to a distributed one is to increase the overall scalability of the system. Hence to have a highly scalable system is desirable in any form of distributed system.

(3) *Transparency:* Another desirable characteristic is to have transparency in the operation. From the user's and the subsystem's point of view, the underlying systems should be transparent. Primarily, transparency can be of two types – *location transparency* and *system transparency*. The first type talks about the need to be transparent regarding the location disparity between different systems. The second talks about the need to be transparent about system issues like failure, concurrency, scaling, migration and so on.

1.2.2 Types of Distributed System

Distributed systems can be divided into mainly three types: distributed computing systems, distributed information systems and distributed pervasive systems. The first type of system is mainly concerned with providing computations in a distributed manner. The second type of system is mainly concerned with providing

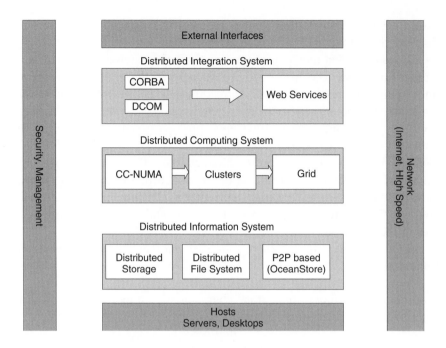

Figure 1.1   Distributed system landscape.

information in a distributed manner, while the third type is the next-generation distributed system, which is ubiquitous in nature.

1.2.2.1   Distributed Computing Systems

Distributed computing systems provide computations in a distributed manner. Computing power is needed in many different industries, including banking and finance, life sciences, manufacturing and so on. If we look at the computing resources available, we shall find that the laptops of today are perhaps as powerful as servers a decade ago. Moore's law, which states that computing power doubles every 18 months, is valid even today and will probably be true for the next 5–6 years. With the growth of the multicore technologies, Moore's law can be extended even further [4]. Computing power is increasing and so is demand. In this rat race, researchers have found an able ally in the form of networking. Between 2001 and 2010, while processing power is supposed to increase 60 times, networking capabilities are supposed to increase by 4000 times. This means that at the same cost, 4000 times the same bandwidth will be available in 2010 as compared to 2001 [5]. Therefore, the computing architectures developed a decade back will probably require a rethink based on the technological progress in the fields of computers and networks. Last decade saw the development of a field called *cluster computing* [6], where different computing resources are connected together using a very-high-speed network like Gigabit Ethernet or more recently Infiniband [7]. In addition to the technological progress and the huge requirement of computing power, the enterprises have also undergone a radical shift in IT operations in the last few years. Enterprises are now witnessing increasing collaboration and data sharing among the different participating entities, resulting in the need for and use of distributed resources and computing. Another important element that has increased the complexity of IT operations is the need for integration of different applications: middlewares developed on different platforms and by different vendors. We are also seeing a spurt of mergers and acquisitions that require integration of technologies across enterprises. Moreover, the enterprises are outsourcing the nonessential elements of the IT infrastructure. The dual pull of requiring more computing power and the integration of heterogeneous components into the IT infrastructure has led to the development of *grid technology*. This technology is seeing a classical evolution pattern. Initiated by the academic and research community to fulfill its needs, it is slowly being adopted by the enterprises, especially those who have high computing needs, such as the life sciences, finance and manufacturing industries. However, the promise of grid computing goes beyond that and the next few years should see a gradual adoption of grid as the natural choice among the other enterprises. But a widespread adoption of grid computing depends upon the ability of researchers and practitioners to reduce the pitfalls that lie along the way. One such pitfall is *security*, which is the focus of this book as a whole. In this chapter we will briefly talk about grid computing's evolution, benefits and concerns.

1.2.2.2 Distributed Information Systems

Distributed information systems are responsible for storing and retrieving information in a distributed manner. There are many manifestations of this type of distributed system. The underlying storage system can be distributed in the form of storage area networks (SANs). SANs have become de facto storage infrastructures in most enterprises. SAN is a high-speed data storage network that connects different types of storage device. One of the most popular modes of storage communication is the Fibre Channel fabric. Another paradigm of the distributed information system is the distributed file system (DFS). The first secure DFS in common use was atheos file system (AFS) [8]. This file system was later followed by DFS [9]. AFS servers store sub-trees of the file system and use Kerberos [10] to provide authenticated access to the trees. Network file system (NFS) is another very popular DFS, which allows users distributed over the network to access distributed files. With the growth of peer-to-peer (P2P) technologies, highly-distributed storage is in vogue. Systems like OceanStore [11] are becoming popular. This uses a large number of untrusted storage devices to store redundant copies of encrypted files and directories in persistent objects. Objects are identified by globally unique identifiers (GUID), which are generated in a similar fashion to the unique identifiers in SAN file system (SFS). Each identifier is a hash of the owner's public key and a name. Objects can point to other objects to enable directories. All objects are encrypted by the client. By replicating the objects among servers, clients can even avoid malicious servers deleting their data. The extensive use of replication and public keys makes revocation of access and deletion of data difficult to achieve, but it does provide a nice model for a completely decentralized DFS.

1.2.2.3 Distributed Integration Systems

Distributed integration systems are responsible for integrating applications, policies and interfaces across diverse distributed systems. The last couple of decades have seen numerous implementations of distributed computing, such as CORBA [12], Java RMI [13], DCOM [14] and so on. None of these systems were taken up in a big way by the industries, mainly because of their tightly-coupled nature. Current trends in the application space suggest that enterprises are moving away from monolithic tightly-coupled systems toward loosely-coupled dynamically-bound components. With the growth of the Internet as a premier means of communication, a new paradigm called the Web Services [15] emerged, facilitating a new style of architecting systems, termed as service-oriented architecture (SOA). Web Services can be thought of as reusable, loosely-coupled software components that are deployed over the network, or specifically the World Wide Web. There are some advantages that the experts claim as the major reasons for the adoption of Web Services as a de facto standard for application integration. These are:

(1) *Simplicity:* Implementation of Web Services is very simple from the point of view of programmers and as a result, easy and fast deployments are possible. All the underlying technologies and protocols are based on Extended Markup Language (XML) [16], which is simple and intuitive.
(2) *Loosely coupled:* Since the very design of Web Services is based on loose coupling of its different components, they can be deployed on demand.
(3) *Platform independent:* Web Services architecture is platform- and language-independent since it is based on XML technologies. Therefore, one can write a client in C++ running on Windows, while the Web Service is written in Java running on Linux.
(4) *Transparent:* Since most of the deployed Web Services use Hypertext Transfer Protocol (HTTP) [17] for transmitting messages, they are transparent to firewalls, which generally allow HTTP to pass through. This may not always be the case for CORBA, RMI and so on.

According to many experts, CORBA and RMI provide a much better alternative to Web Services because of the flexibility and features that CORBA provide. Moreover, performance-wise the CORBA/RMI combination may be better than protocol designed over HTTP. However, because of its simplicity and the backing of the big commercial vendors, Web Services is steadily becoming a standard which none can ignore. There are many forums where debates are being pursued as we move on to the different components which constitute the Web Services. There are three main components of Web Services:

- *SOAP:* The Simple Object Access Protocol (SOAP) [18] is a lightweight protocol for exchange of information between diverse and distributed computing environments. It combines the extensibility and portability of XML with the ubiquitous Web technology of HTTP. It provides a framework for defining how an XML message is structured, using rich semantics for indicating encoding style, array structure and data types.
- *WSDL:* The Web Service Description Language (WSDL) [19] can be used to describe a Web Service, providing a standard interface. A WSDL document is written in XML and describes a service as a set of endpoints, each consisting of a collection of operations. XML input and output messages are defined for each operation and their structure and data types are described using an XML Schema in the WSDL document. The Web Description Services Language (WDSL) and XML Schema provide a complete definition for the service interface, allowing programmatic access to the Web Service in the manner of an API. Tasks like data requests or code executions can be performed by sending or receiving XML messages using, for example, SOAP.
- *UDDI:* The Universal Description, Discovery and Integration (UDDI) [20] specification defines a way to publish and discover information about Web Services. It is a collaboration between Ariba, IBM and Microsoft to speed interoperability and adoption of Web Services. The project includes a business registry (an XML document) and a set of operations on it. The registry can be used by programs to find and get information about Web Services and check compatibility with them, based on their descriptions. UDDI allows categorization of Web Services so that they can be located and discovered, and WSDL enables a programmatic interface to a service once it has been located.

1.2.3 Different Distributed Architectures

There are four different types of architecture that are used for designing distributed systems, namely client–server-based systems, Multinode systems, P2P systems and service-oriented systems. The first type of system is a client- and a server-based system, the second type of system distributes the data or the information across multiple nodes or systems, the third type of system is a P2P-based architecture where all components are peers or at the same level, and the last type of system is a federated model where interactions happen via standards-based messages.

1.2.3.1 Client–Server-Based Architecture

Client–server-based architecture is the most popular distributed system that has been used over the years. In this architecture, the server is responsible for providing service to the client or a set of clients. In a client–server kind of environment, a client requests a service from the server, which the server provides over the network in a remote environment. The main advantage of a client–server system is that the business services are abstracted from the set of clients accessing them. Security is implemented at the link and the end server, while fault tolerance is applied at the server end by replicating the functionality. Though extremely popular, there are some inherent limitations in this type of architecture, which led practitioners and researchers to other models.

- *Scalability:* One of the primary limitations of this model is scalability. If the number of users increases significantly, the system fails to handle such a large load. There are two ways to handle this issue: scale up or scale out. Scaling up means moving to a higher end server to handle the same type of request. Though this may be an effective solution in some cases, it does not scale as there is a limitation to scaling up. The second approach, or scale-out approach, distributes the server into multiple servers, which improves scalability. We will talk about this approach later.
- *Flexibility:* Just having a client and a server reduces the overall flexibility of the system, the reason being that database management, application business logic and other processes are embedded in the server code and hence inflexible. Practitioners and designers slowly moved to a three-tier architecture mainly to tackle this problem.

1.2.3.2 Multinode

One variation of the client–server technology distributes the server into multiple nodes that can be used for parallel processing. There are several advantages of such a multinode configuration, namely *performance*, *fault tolerance* and *scalability*. Performance can be improved, since the different nodes involved in the process provide part of the service a single node was supposed to perform. Different components of the multinode system are: processing nodes, scheduler or load balancer and clients. Having different nodes perform similar actions can result in improvement of fault tolerance. Moreover, multinode systems improve scalability since they can scale out instead of scaling up. However, the advantages of multinode systems come at a cost, which is complexity. Managing synchronization, security, load balancing and so on in such an environment is extremely challenging.

1.2.3.3 Peer-to-Peer

The third type of architecture, which is becoming at the moment, is P2P. This type of system is different from client–server-based systems as, in P2P systems, all the nodes in the distributed system participate in the same hierarchy. This means that there is no concept of client and server, and each participant assumes the role of client and server based on need and requirement. Systems like Gnutella, Napster and so on are based on such principles. P2P systems have found significant applications in the area of file distribution and transfer. They are also being applied in the area of data and information storage. P2P systems have several advantages in terms of scalability and fault tolerance. Since the system is dependent on end systems and nodes, it scales infinitely. The scalability property is exhibited by all the P2P systems. Similarly, fault tolerance is also an important characteristic of such a system. However, several challenges exist, in the form of security and service level agreement (SLA). Since the end systems are responsible for performance, guaranteeing service is almost impossible. Management of security is also extremely difficult as maintaining security at the end systems is a challenge.

1.2.3.4 Service-Oriented Architecture

SOA is the latest in the evolution of distributed architectures, which builds upon the client–server and other such distributed architecture models. SOA implementations revolve around the basic idea of a service. A service refers to a modular, self-contained piece of software, which has a well-defined functionality expressed in abstract terms independent of the underlying implementation. Basically, any implementation of SOA has three fundamental roles: service provider, service requestor and service registry, and three fundamental operations: publish, find and bind. The service provider publishes details pertaining to service invocation with a services registry. The service requestor finds the details of a service from the service registry. The service requestor then invokes (binds) the service on the service provider. Web Services, described earlier, represent the most popular form of implementation of SOA.

1.2.4 Challenges in Designing Distributed Systems

The challenges in designing a distributed system lie in managing the different disparate entities responsible for providing the end service. Synchronization between the different entities needs to be handled. Similarly, security and fault tolerance are extremely important and need to be handled as well.

1.2.4.1 Synchronization

One of the most complex and well-studied problem in the area of distributed systems is synchronization. The problem of synchronizing concurrent events also occurs in nondistributed systems. However, in distributed systems, the problem gets amplified many times. Absence of a globally-shared clock, absence of global shared memory in most cases and the presence of partial failures makes synchronization a complex problem to deal with. There are several issues, like clock synchronization, leader election, collecting global states, mutual exclusion and distributed transactions, which are critical and have been studied in detail in literature.

- *Clock synchronization:* Time plays a crucial role as it is sometimes necessary to execute a given action at a given time, timestamping data/objects so that all machines or nodes see the same global state. Several algorithms for clock synchronization have been proposed, which include synchronization of all clocks with a central clock or through agreement. In the first case, the time server or external clock periodically sends the clock information to all the nodes, either through a broadcast or through multicast mechanisms, and the nodes adjust the clock based on the received information and the round-trip time calculation. In the second mechanism, the nodes exchange information so that the time clock can be calculated in a P2P fashion. It is to be noted that clock synchronization is a major issue in distributed systems and clock skew always needs to be considered when designing such a system.
- *Leader election:* This is another critical synchronization problem used in many distributed systems. Many varieties of solution are available, ranging from the old leader forcing the new leader on the group members based on certain selection criteria, to polls or votes where the node receiving the maximum number of votes gets elected as the leader.
- *Collection global state:* In some applications, especially when debugging a distributed system, knowledge of the global states is especially useful. Global state in a distributed system is defined as the sum of the local states and states in transit. One mechanism is to obtain a distributed snapshot which represents the consistent and global state in which the distributed system would have been. There are several challenges in moving a process to the consistent state.
- *Mutual exclusion:* In some cases, it is required that certain processes access critical sections or data in a mutually-exclusive manner. One way to tackle such a problem is to emulate the centralized system by having the server manage the process lock through the use of tokens. Tokens can also be managed in a distributed manner using a ring or a P2P system, which increases the complexity.

1.2.4.2 Fault Tolerance

If we look at the issue of fault tolerance from the distributed systems perspective, it is both an opportunity and a threat. It is an opportunity as distributed systems bring with them natural redundancy, which can be used to provide fault tolerance. However, it is a threat as the issue of fault tolerance is complex, and extensive research has been carried out in this area to tackle the problem effectively. One of the issues that haunts distributed systems designers is the source of many failures. Failures can happen in processing nodes and transmission media, and due to distributed agreement.

- *Processing sites:* The fact that the processing sites of a distributed system are independent of each other means that they are independent points of failure. While this is an advantage from the viewpoint of the user of the system, it presents a complex problem for developers. In a centralized system, the failure of a processing site implies the failure of all the software as well. In contrast, in a fault-tolerant distributed system, a processing site failure means that the software on the remaining sites needs to detect and handle that failure in some way. This may involve redistributing the functionality from the failed site to other, operational, sites, or it may mean switching to some emergency mode of operation.
- *Communication media:* Another kind of failure that is inherent in most distributed systems comes from the communication medium. The most obvious, of course, is a permanent hard failure of the entire medium, which makes communication between processing sites impossible. In the most severe cases, this type of failure can lead to partitioning of the system into multiple parts that are completely isolated from each other. The danger here is that the different parts will undertake activities that conflict with each other. Intermittent failures are more difficult to detect and correct, especially if the media is wireless in nature.
- *Errors due to transmission delays:* There are two different types of problem caused by message delays. One type results from variable delays (jitter). That is, the time it takes for a message to reach its destination may vary significantly. The delays depend on a number of factors, such as the route taken through the communication medium, congestion in the medium, congestion at the processing sites (e.g. a busy receiver), intermittent hardware failures and so on. If the transmission delay is constant then we can much more easily assess when a message has been lost. For this reason, some communication networks are designed as synchronous networks, so that delay values are fixed and known in advance. However, even if the transmission delay is constant, there is still the problem of out-of-date information. Since messages are used to convey information about state changes between components of the distributed system, if

the delays experienced are greater than the time required to change from one state to the next, the information in these messages will be out of date. This can have major repercussions that can lead to unstable systems. Just imagine trying to drive a car if visual input to the driver were delayed by several seconds.

- *Distributed agreement:* The problem of distributed agreement has been briefly touched upon in the previous subsection. There are many variations of this problem, including time synchronization, consistent distributed state, distributed mutual exclusion, distributed transaction commit, distributed termination, distributed election and so on. However, all of these reduce to the common problem of reaching agreement in a distributed environment in the presence of failures.

1.2.4.3   Security

Perhaps the most compelling challenge associated with distributed systems is the issue of security. The complexity of the issue arises from the different points of vulnerability that exist in a distributed system. The processing nodes, transmission media and clients are the obvious points that need to be secured. With the growth of heterogeneity in different layers of enterprise infrastructure, the complexity increases enormously. This whole book is devoted to this subject. In the next section, we will provide a brief motivation for different layers of distributed systems in an enterprise scenario and touch upon the security issues to be delved into in this book.

1.3   Distributed Systems Security

As mentioned earlier, security in distributed systems is critical and absolutely essential. However, it is also extremely challenging. Distributed security in the digital world is akin to security in the real world. As the last few years would suggest, protecting physical infrastructure is turning out to be a nightmare for security professionals. The reason is that malicious adversaries can reside anywhere, and everything is their potential target. In the digital world as well, protecting the infrastructure is turning out to be a catching game. The main reason for this is that the IT infrastructure in all enterprises is distributed in nature. Before understanding the security in distributed systems in relation to enterprise IT, we need to understand the enterprise IT landscape. In this section, we will discuss the enterprise IT scenario in a layered perspective. The whole book will then be aligned to this layered view with respect to distributed IT security.

1.3.1   Enterprise IT – A Layered View

Figure 1.2 shows a high-level view of the layered enterprise. The view consists of four main layers: hosts, infrastructure, applications and services. While the

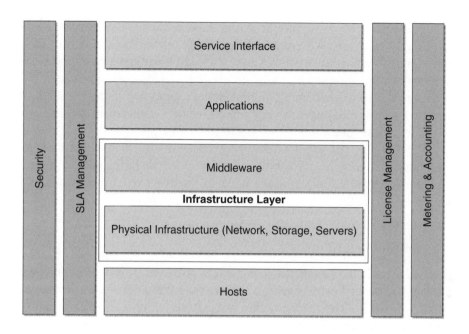

Figure 1.2 Layered enterprise view.

host layer consists of client desktops and low-end servers, forming the lowest stratum of the enterprise IT, the infrastructure layer consists of network, storage and middleware functionalities, which are used by the application, host and service layers. The applications are custom and component of the shelf (COTS) applications that are used in the enterprises in a day-to-day manner. Finally, we have the service layer, which provides standards-based services to the external world as well as to the enterprise itself. We will take this view into account for all our subsequent discussions.

1.3.1.1 Hosts

The lowest layer in the enterprise consists of hosts, which are mainly composed of client desktops and low-end servers. Even a few years back, hosts were meant only to submit requests to servers and perform some low-end user-level tasks like editing Word files and so on. However, with the growth of grid computing, concepts like cycle stealing, scavenging and so on are coming to the fore. Middleware technologies are able to take advantages of idle desktops or low-end servers like the blades for distributed processing. With the growth of P2P technologies, hosts are also managing distributed files in a much more scalable manner. Technologies like Torrent [21] are redefining the way storage is carried out today. With the growth of dimensions of hosts, several issues need to be tackled which were not a problem before.

- *Manageability:* The issue of managing heterogeneous systems is becoming increasingly complex. With hundreds and thousands of hosts a part of the computing and storage infrastructure, this is an administrator's nightmare. Several management and monitoring tools are needed to address this problem.
- *Metering:* With hosts becoming more and more important in the overall business scenario, metering assumes a very important role. How to meter and what to meter are serious questions.
- *Security:* Perhaps the most challenging issue in the host-based storage and computation is security. Not only do hosts need to be protected from malicious outside agents, infrastructure needs to be protected from malicious hosts as well.

1.3.1.2 Infrastructure

The second layer in the IT enterprise is the infrastructure. It is diverse and complex because of the sheer heterogeneity of products, technologies and protocols used in enterprises today. The infrastructure is basically composed of two main components: the physical infrastructure, consisting of high-end servers, storage and networks, and the middleware, consisting of the cluster and grid middlewares.

- *Physical infrastructure:* Physical infrastructure consists of the server infrastructure, network infrastructure and storage infrastructure. One of the key characteristics of the physical infrastructure is heterogeneity. From the size and type of servers used, through the networking speed to the storage devices, heterogeneity remains a key ingredient. Another key characteristic is that each component of the physical infrastructure is distributed in nature, making security a major concern.
- *Middleware:* Cluster and grid middleware dominates in a high-performance environment. Integration of grid technologies with mainstream IT is a challenge.

1.3.1.3 Applications

Applications address the diverse enterprise needs of an enterprise, be they business applications or horizontal technological applications. Of special importance is the emergence of Web-based applications as a crucial component of enterprise landscape, essential for delivering the right functions to consumers via the Web. From a security perspective, application security has assumed major importance in the recent past. Security issues may crop up due to either weakness in the design of an application, or an insecure coding practice, which can compromise some of the security requirements. In the case of the Web, the openness of the medium and the protocols is responsible for further security complexity.

1.3.1.4 Services

Services represent a higher level of interaction in distributed systems, building over the underlying applications and data. Hence the typical underlying distributed system security issues (including confidentiality, integrity and so on) are applicable; additionally, the specific concerns arising out of the loose coupling and interaction via XML messages introduce extra complexities. The higher level of loose coupling required for SOA mandates more flexible ways of handling security for SOA. Additionally, standards will be key as there is a need to interoperate across heterogeneous implementations of underlying systems. Finally, the openness and plain-text nature of XML-based distributed invocations is a cause of further complexity and higher vulnerability. Likewise, typical distributed-system attacks like DOS, cross-site scripting attacks and so on manifest at service level too, albeit with variations.

1.3.2 Trends in IT Security

As we move toward a distributed IT infrastructure, security issues become more and more critical. The pervasive growth of the IT infrastructure, along with its heterogeneous nature, makes security a really complex issue to look at. If we look at a typical IT infrastructure, there are hundreds of different applications that are interacting with one another. The applications are either custom built, vendor products or even open-source systems. Each of the products interacts with complex sets of infrastructure components, including servers, desktops, middlewares and so on. Added to this complexity is that of the heterogeneous networking infrastructure, including wired, wireless and so on, and devices like BlackBerrys, personal digital assistants (PDAs) and others. With the growth of sensor networks, integration of IT infrastructure and small sensor motes will make the problems exceedingly challenging. With the heterogeneity and pervasive nature of enterprises set to grow, several security trends have been identified in this section, which are slowly being adopted by enterprises around the world. The key security trends that can be observed are: movement of security to higher layers, protection of the periphery, protection of identities, standardization and integration of heterogeneous policies and infrastructure.

1.3.2.1 Security in Higher Layers

One of the security trends that is observed currently is the movement of security implementation to higher layers. If we look at the different layers of enterprise systems, security protocols and systems are available at each and every one. For example, most of the enterprises conform to SOA. Different security protocols are available at the infrastructure layer, the middleware layer and so on. Enterprises are slowly exploring the ideas of having security at the Web Services layer, which has led to the standardization and development of WS-Security standards. Similarly, enterprises are looking at securing the higher layers so that more flexibility can be obtained. However, one of the issues in moving security up the layers is performance versus scalability. The higher the security implementation the more the security overhead, and hence the more the performance overhead. Therefore, the decision to have security at a particular layer depends on the amount of flexibility that the system requires and the performance requirement of the system. Taking the above example, instead of WS-Security, one can implement Transport Layer Security (TLS). The performance of a TLS-based system will be more than that of a WS-Security-based system; however WS-Security provides an end-to-end secure environment which is not provided by TLS.

1.3.2.2   Protection of the Periphery

Another important trend that can be observed in enterprise-security cenarios is that the security is provided at the periphery to protect the network by filtering the traffic flowing to and from it. Different types of filtering technique are employed to protect the data flowing into the network. Filtering can be as simple as going through the packets and preventing data coming from certain ports. Similarly, requests going to a particular port can be prevented. However, enterprises are also moving toward more sophisticated methods of filtering, like application-level filtering and XML-based filtering techniques. In these techniques, the filters or firewalls actually look into the XML or application payload and identify whether the packet is of a malicious nature.

1.3.2.3   Protection of Identities

When I reflect upon my activities today, I find that I have used multiple credentials to access resources of different forms. I used my company identity card to enter the office premises, entered the password to get into the office network, used my smart card to access the high-security lab, used my personal identification number (PIN) to access my ATM account, and used my passport to get a US visa – and this was just one day. Different identity checks were required by different systems, and my identities were in different forms, which I either carried in my head or as a card or a paper. I am surely not an exception; every one of us is doing the same, maintaining multiple credentials to access different forms of resource. This has really become pronounced with the growth of Information Technologies, where there are multitudes of system interfaces which require some sort of user authentication. As a result, individuals possess multiple digital identities and credentials, many of which are short-lived. At this point, one may be concerned about the relationship between identities and credentials. The identity of an individual user is unique; however, it may be manifested in different ways to disparate systems through user credentials. For example, my identity credential

to the United States consulate is my passport, while to the company network
it is the combination of the network's user ID and password. Therefore, when
we talk of managing different user identities, it is actually the user-identification
credentials we are talking about. However, credentials go beyond just identifying
the user: they may authorize a user to access a certain resource or be used as a
proof of authentication. Credentials can be short-lived, for example identity cards
or passwords which expire when the individual leaves a company, or after a fixed
amount of time. Other examples of short-time credentials are the tickets issued in
busses for a short ride. Individuals manage their credentials by a combination of
papers, cards and their own memory, as I did today. Secure management of user
credentials is a very important challenge. Identity theft topped the list of complaints to the United States Federal Trade Commission in 2002, accounting for
43% of all complaints [22]. Therefore, identity and user-credential management
is surely a very important problem, and several research and development efforts
are being undertaken in this direction.

1.3.2.4   Standardization

With the growth of heterogeneity in the enterprises, standardization is fast becoming a key for any enterprise security system. If we look at any enterprise, the
number of heterogeneous elements available is mind-boggling. Several enterprises
over the years have custom-built applications, even middlewares, to interact with
vendor products. When architects look at the security issues in such enterprises,
they find the need to integrate security across these products, middlewares and
applications. The way to solve the problem is through standardized interfaces
and protocols. There are a couple of advantages to taking the standardized route,
especially in designing the security systems in enterprises. Firstly, rather than
designing a custom protocol, standards are based on well-established theoretical
bases and principles. Hence, through standards, one can be sure that vulnerabilities
are not introduced in those layers. Secondly, standard interfaces make integration
a slightly less cumbersome problem.

1.3.2.5   Integration

Perhaps the most complex and challenging problem in any enterprise is the integration of different protocols, standards, applications, middlewares and so on.
This becomes especially complex for new and evolving technologies, like grid
computing for example. Though technically grid computing is a powerful technology which provides flexibility and performance in the current infrastructure setup,
when enterprises move into the grid environment, the challenges of integration
just hide all the benefits that exist. Enterprises which have application servers,
data-base tiers, different business intelligence products and monitoring tools, and
management systems, would integrate with an open-source Globus toolkit that

is based on standards like Security Assertion Markup Language (SAML) and WS-Security. However, the enterprises do not support those standards and either the enterprise applications have to move to the newer standards, which may involve a lot of work and customization, or the grid systems must be customized to work with the existing standards and protocols. In most cases, enterprises prefer the second route as they generally do not want to touch the systems which 'work'. As a result, the vicious circle of newer technologies and integration difficulties persists.

1.4 About the Book

In this book we look at the global picture of the distributed computing systems and their security vulnerabilities, and the issues and current solutions therein. We divide the distributed systems into four layers: infrastructure, host, application and service. The reason for this layering lies in the fact that enterprises have systems built in this manner, and integration issues come to the fore when analyzing them as we have done. The host issues look at the issues pertaining to a host in a distributed computing system and the vulnerabilities that it will be subjected to; vulnerabilities include mobile codes coming into the system and tasks being executed. The infrastructure level issues concern the infrastructure as a whole, that is the networking and the mobile infrastructure on which the distributed systems are perched. The application layer is concerned with applications that are being developed on top of the infrastructure. Lastly, the service layer looks at building distributed services. As we can see, each of the layers presents unique challenges in terms of security. Moreover, the book looks at the orchestration of applications and services across the different layers in order to look at the global picture.

1.4.1 Target Audience

The book does not assume that the reader is an expert in security or distributed systems technologies. However, some prior knowledge about general security principles and/or distributed computing technologies will be required to understand the chapters covering advanced security issues. The book is primarily targeted at architects who design and build secure distributed systems. It would also benefit managers who make business decisions and researchers who can find research gaps in existing systems.

- *Professionals and architects:* Through this book, professionals and architects working on distributed systems will be made aware of the security requirements. It will also enlighten them about the security features of some existing open-source as well as proprietary products. The book also aims at identifying processes and models which could help architects design more secure systems.

- *Managers and CIOs:* Though the book has significant technical depth, managers and CIOs will gain significantly from it by understanding the processes, gaps and solutions which exist. The book therefore will be able to provide them with information which will be useful for making important business decisions.
- *Researchers and students:* Experienced researchers and students in the field of distributed computing will be able to get a comprehensive overview of all the security issues in distributed computing. This will help them make important research decisions by analyzing the existing gaps.

References

[1] Isaac Asimov Online (2008) http://www.asimovonline.com, accessed on June 13th, 2008.
[2] Orkut (2008) http://www.orkut.com, accessed on June 13th, 2008.
[3] Facebook (2008) http://www.facebook.com, accessed on June 13th, 2008.
[4] Koch, G. (2005) *Discovering Multi-Core: Extending the Benefits of Moore's Law*, Technology Intel® Magazine.
[5] Vilett, C. (2001) *Moore's Law vs. Storage Improvements vs. Optical Improvements*, Scientific American.
[6] Rajkumar B. (eds) (2003) *High Performance Cluster Computing: Architectures and Systems*, Kluwer Academic Publishers.
[7] InfiniBand Trade Association (2008) *InfiniBand Architecture Specification*, Vol. **1**, Release 1.1, November 2002. Available from http://www.infinibandta.org, accessed on June 13th, 2008.
[8] Howard, J.H., Kazar, M.L., Menees, S.G. *et al.* (1988) *Scale and Performance in a Distributed File System*, Vol. **6.1**, ACM Transactions on Computer Systems, pp. 51–81.
[9] Kazar, M.L., Leverett, B.W., Anderson, O.T. *et al.* (1990) Decorum file system architectural overview. *Proceedings of the Usenix Summer 1990 Technical Conference*, USENIX, pp. 151–64.
[10] Neumann, C. and Ts'o, T. (1994) Kerberos: an authentication service for computer networks. *IEEE Communications Magazine*, 32(9):33–38, September 1994.
[11] Kubiatowicz, J., Bindel, D., Chen, Y. *et al.* (1999) *Oceanstore: An Architecture for Global-Scale Persistent Storage*, ASPLOS.
[12] Vinoski, S. (1997) CORBA: integrating diverse applications within distributed heterogeneous environments. *IEEE Communications Magazine*, 14, 2 February 1997.
[13] Grosso, W. (2001) *Java RMI*, O'Reiley.
[14] Ruben, W. and Brain, M. (1999) *Understanding DCOM*, Prentice Hall. ISBN 0-13-095966-9.
[15] Cerami, E. *Web Services Essentials*, O'Reilley, ISBN: 0596002246, 2002.
[16] Ray, E. *Learning XML*, O'Reilley, ISBN: 0596004206, 2003.
[17] Fielding, R., Gettys, J., Mogul, J. *et al.* (1999) Hypertext Transfer Protocol – HTTP 1.1, IETF RFC 2616, June, 1999.
[18] W3C Team (2003) SOAP Version 1.2, Part 0: Primer, W3C Recommendations, June 2003.
[19] W3C Team (2001) Web Services Description Language (WSDL) 1.1, W3C Note, March 2001.
[20] Computer Associates (2005) IBM, Microsoft, Oracle, SAP, SeeBeyond Technologies, Systinet, and Others, UDDI v.3.0, *OASIS Standard*, Feb 2005.
[21] BitTorrent (2008) http://www.bittorrent.org, accessed on 13th June 2008.
[22] Basney, J., Yurcik, W., Bonilla, R. and Slagell, A. (2006) Credential Wallets: A Classification of Credential Repositories, Highlighting MyProxy. 31st Annual TPRC, Research Conference on Communication, Information, and Internet Policy, Sep. 2006.

2

Security Engineering

2.1 Introduction

Security engineering as a subject is essentially about the activities involved in engineering a software application that is secure and reliable and has safeguards built in against security vulnerabilities. It is a very vast subject that covers a wide range of topics, from technologies addressing specific security requirements such as authentication, authorization, confidentiality and nonrepudiation, to security architectural solutions, patterns and design principles, and a plethora of other activities that are required to develop secure software.

Many of these technologies are well understood and well covered in several security engineering books. However, the often ignored, but equally (if not more) important dimension of security engineering is about how to engineer an application that is secure by specification, design and development. The fundamental reason for the existence of so many security vulnerabilities is defective software. For many an enterprise, developing business functionality takes precedence over almost everything else; as a result, security is often considered something to be plugged in as an after thought. The prevalent industry approach to security seems to be 'patch what is found to be breakable'. Security is rarely considered by software development firms as an integral part of the software development lifecycle (SDLC). As a result, security cross-checks are almost never defined as a part of the SDLC. Moreover, software engineers are typically not aware of or trained on common security vulnerabilities and the techniques to guard against them.

Defective software causes several billions of US dollars' worth of financial losses every year across the globe [1]. Fixing a security bug only after a vulnerability has been reported and/or exploited is the least desirable situation and is often very expensive in terms of financial loss and brand loss. Organizations are increasingly realizing that it is much cost-effective if the security vulnerabilities can be prevented from appearing in the software in first place. Finding and

Distributed Systems Security   A. Belapurkar, A. Chakrabarti, S. Padmanabhuni, H. Ponnapalli, N. Varadarajan and S. Sundarrajan © 2009 John Wiley & Sons, Ltd

fixing security vulnerabilities in the development phase is the best possible way of addressing security. Security is best achieved when it is considered at every step of the software development process. In that sense, security is just like software quality – it has to be built in rather than added on as an extra layer.

Incorporating aspects of security into the SDLC is still evolving as a concept, although there are a few documented processes and process models available to practitioners [2–7]. As a matter of fact, many organizations have started taking this seriously fairly recently. For example, Microsoft took a strategic decision to start addressing security as part of its development culture in the year 2002, under the broad umbrella of what it calls Trustworthy Development [8]. This has helped it make very good progress in reducing the number of security bulletins it needs to issue on its products. Process models such as the systems security engineering capability maturity model (SSE-CMM) [2] have been around for some time; however, the adoption of most of them has been very minimal until recently.

In this chapter, we will cover in depth this often ignored subject of addressing security as an integral part of the SDLC. We will provide an overview of some of the prevailing secure development lifecycle processes. We will also cover a few common minimal security engineering activities like security requirements, threat modeling, security architectures and code reviews, and security testing. This topic is not covered in details in most books, with a notable exception of a couple [8, 9] whose focus is primarily on security in the software development lifecycle.

2.2 Secure Development Lifecycle Processes – An Overview

There are very few software development processes and process models available in the industry today in which security is considered from the ground up as an integral part of the entire process. Though many of the widely popular process models like the capability maturity model (CMM) guarantee good-quality software, security is not explicitly considered in them and hence they are not directly amenable for use in developing secure software.

While a process model defines a set of best practices and provides a generic architecture (high-level logical process areas), it does not define any specific process. This means several processes can be defined following the same guiding principles, as established by a single process model. A process is an instance of a process model template which provides a specific set of activities, and details on how those activities are to be carried out.

This section provides details on some of the prevalent process models as well as a few proven industry processes. However, a comprehensive survey of the secure development processes is out of the scope of this book. We will cover SSE-CMM [2], Microsoft's Security Development Lifecycle (SDL) [8], the comprehensive lightweight application security process (CLASP) [3] and the US Department of

Homeland Security (DHS)'s Build Security In (BSI) security assurance initiative [5]. There are other secure development lifecycle processes, like trusted capability maturity model/trusted software methodology (T-CMM/TSM), team software process (TSP), information security system development lifecycle from NIST, IEEE P1074, common criteria [10] and so on. A detailed coverage of all these methods and initiatives is out of our scope. Interested readers may refer to [6, 7] for a very detailed coverage of the state of the art in software security assurance.

2.2.1 Systems Security Engineering Capability Maturity Model (SSE-CMM)

SSE-CMM is a process reference model that focuses on implementing information technology systems securely. The SSE-CMM is a community-owned model, and is a result of the dedicated work of many individuals. Being a process reference model, it does not specify any specific set of processes or methodologies to be followed to develop secure software; rather, it expects that organizations will use their existing processes based on some security guidance. In other words, it defines WHAT activities are to be performed rather than HOW they are to be performed. The scope of SSE-CMM includes security engineering activities to develop secure software, including: requirements analysis, design, development, integration, installation, maintenance and decommissioning. The objective of SSE-CMM is to advance security engineering as a well-defined, mature and measurable discipline that enables organizations to make focused security investments in tools/training, and capability-based assurance.

SSE-CMM categorizes security engineering into three process areas: risk, engineering and assurance. The risk process area identifies, prioritizes and helps in managing the risks associated with the system. The engineering process area deals with the implementation of solutions and countermeasures to minimize the risks identified by the risk process area. The assurance process area deals with providing the necessary confidence level in the security solutions implemented for/by the customers.

Further, SSE-CMM practices are divided into two dimensions: domain and capability. Domain practices are the best practices in security engineering; they are also called base practices. Capability practices pertain to process management and institutionalization capability; they are referred to as generic practices. Generic practices represent the activities to be undertaken as part of carrying out a base practice. Putting the base practices and generic practices together helps check an organization's capability in performing a particular activity.

SSE-CMM contains 129 base practices (categorized into 22 process areas), of which 61 base practices (organized into 11 process areas) cover all the security engineering activities. The rest cover project and organization domains that have been drawn from the Systems Engineering and Software CMM. The 11 security engineering-related process areas in base practices are as follows:

(1) administer security controls
(2) assess impact
(3) assess security risk
(4) assess threat
(5) assess vulnerability
(6) build assurance argument
(7) coordinate security
(8) monitor security posture
(9) provide security input
(10) specify security needs
(11) verify and validate security

Generic practices are activities that apply to all processes. Generic practices are grouped into logical areas called 'common features', which are organized into five 'capability levels' that represent increasing levels of organizational capability. Unlike the base practices of the domain dimension, the generic practices of the capability dimension are ordered according to maturity. Figure 2.1 provides a depiction of SSE-CMM process areas.

A detailed coverage of these different process areas and their related activities is out of the scope of this book. Interested readers may refer to [2]. SSE-CMM provides a comprehensive framework that gives best practices to develop and maintain secure software.

2.2.2 Microsoft's Security Development Lifecycle (SDL)

The trustworthy computing Security Development Lifecycle is a process adopted by Microsoft for the development of secure software. The process comprises a series of security activities for and deliverables to each phase of Microsoft's development lifecycle. These cover both security engineering and assurance activities. The Microsoft SDL has evolved over the last five years, since it was begun in 2002 to show Microsoft's commitment to improving the security of its products. According to Microsoft, any product that has significant security risk must undergo the SDL. Applications with significant risk include any application that processes sensitive or personal data, is connected to the Internet or used in a networked environment, processes financial data, or is used in an enterprise or other organization.

The SDL involves modifying the organization's development process by integrating a set of well-defined security checkpoints and deliverables. The SDL covers all phases of Microsoft's development processes, including: requirements, design, development, verification, release and response. During the requirements phase, security objectives are clearly specified and the security feature requirements captured. During design, threat models are developed to identify all possible threats to the application in a structured manner. Focused security architecture and

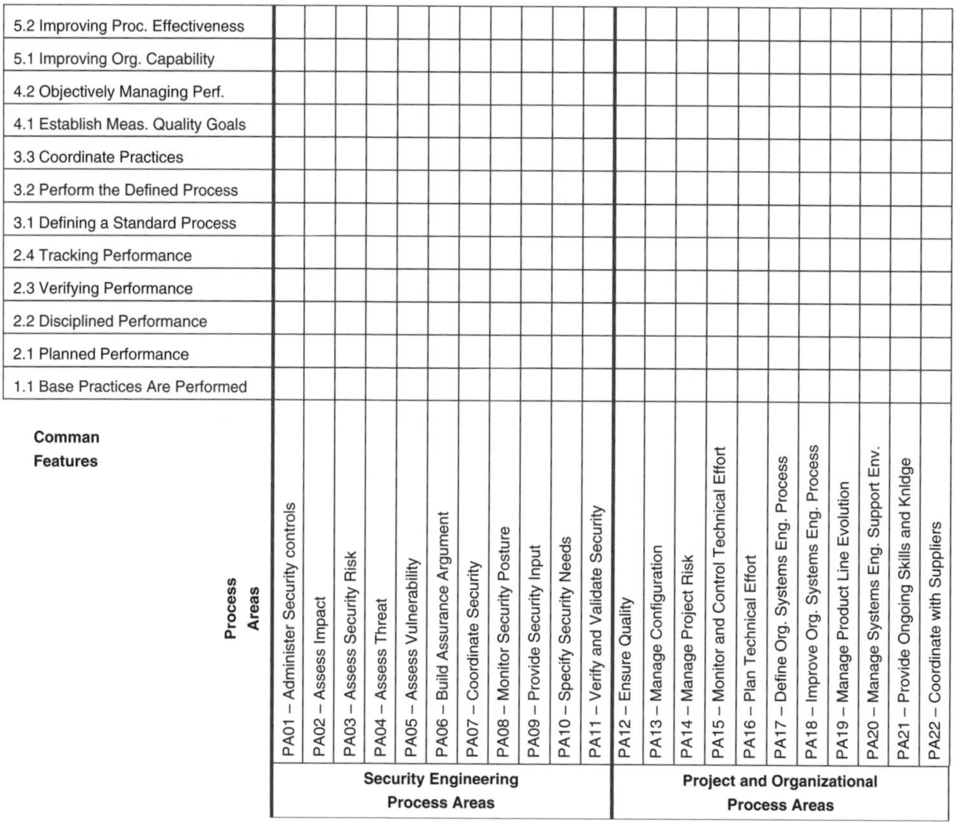

Common Features		Process Areas
5.2 Improving Proc. Effectiveness		
5.1 Improving Org. Capability		
4.2 Objectively Managing Perf.		
4.1 Establish Meas. Quality Goals		
3.3 Coordinate Practices		
3.2 Perform the Defined Process		
3.1 Defining a Standard Process		
2.4 Tracking Performance		
2.3 Verifying Performance		
2.2 Disciplined Performance		
2.1 Planned Performance		
1.1 Base Practices Are Performed		

Security Engineering Process Areas:
- PA01 – Administer Security controls
- PA02 – Assess Impact
- PA03 – Assess Security Risk
- PA04 – Assess Threat
- PA05 – Assess Vulnerability
- PA06 – Build Assurance Argument
- PA07 – Coordinate Security
- PA08 – Monitor Security Posture
- PA09 – Provide Security Input
- PA10 – Specify Security Needs
- PA11 – Verify and Validate Security

Project and Organizational Process Areas:
- PA12 – Ensure Quality
- PA13 – Manage Configuration
- PA14 – Manage Project Risk
- PA15 – Monitor and Control Technical Effort
- PA16 – Plan Technical Effort
- PA17 – Define Org. Systems Eng. Process
- PA18 – Improve Org. Systems Eng. Process
- PA19 – Manage Product Line Evolution
- PA20 – Manage Systems Eng. Support Env.
- PA21 – Provide Ongoing Skills and Knldge
- PA22 – Coordinate with Suppliers

Figure 2.1 SSE-CMM process areas and common features (source: SSE-CMM Ver 3.0).

design reviews form a key checkpoint during the design phase. During development, secure coding guidelines are followed to develop secure software. Static code analysis tools and fuzz testing techniques are used to verify the security of software developed. Code reviews are carried out, with specific focus on security. During the security push (verification) penetration testing, fuzz testing techniques are used to analyze the security posture of the application from a black-box perspective. Before the software is released, a final security review is carried out by a team independent of and different from the development team. The post-release activities are supported through the Microsoft Security Response Center, which analyzes any identified security vulnerabilities and releases security bulletins and security updates. A central security team coordinates the entire gamut of security activities and a security buddy is associated with every project to help it follow the SDL. The security buddy is identified right at the beginning (during the requirements phase) and stays associated with the project till the final security review.

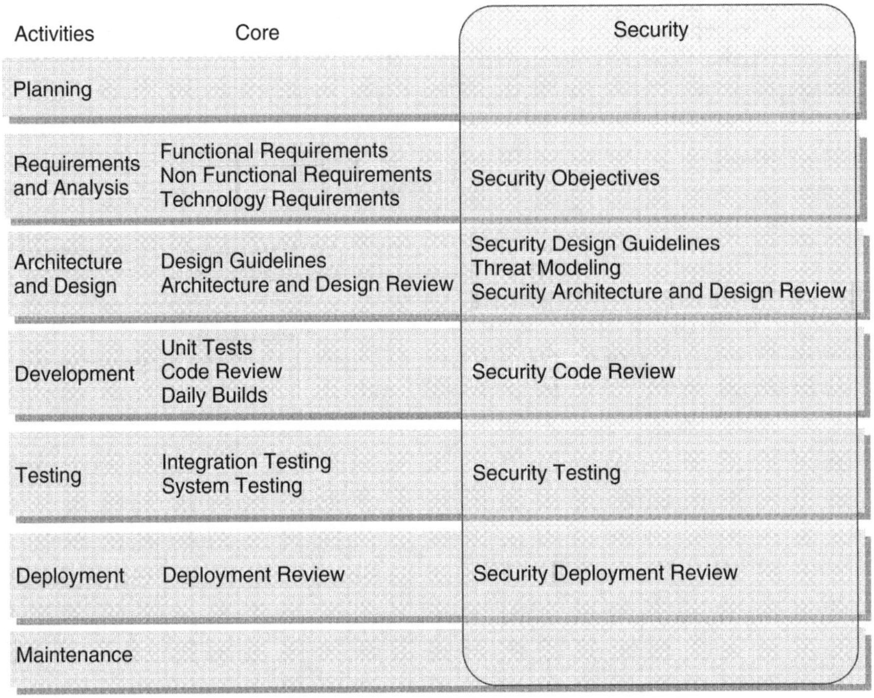

Activities	Core	Security
Planning		
Requirements and Analysis	Functional Requirements Non Functional Requirements Technology Requirements	Security Obejectives
Architecture and Design	Design Guidelines Architecture and Design Review	Security Design Guidelines Threat Modeling Security Architecture and Design Review
Development	Unit Tests Code Review Daily Builds	Security Code Review
Testing	Integration Testing System Testing	Security Testing
Deployment	Deployment Review	Security Deployment Review
Maintenance		

Figure 2.2 Microsoft SDL activities (source: Microsoft Security Engineering Explained).

Figure 2.2 shows a simplified security overlay from Microsoft, depicting typical security engineering activities and the places where they are best fit into an application development lifecycle.

Though the process looks like a typical waterfall, many of the activities are performed repetitively; thus, in reality, it is a spiral model.

Measuring the usefulness and effectiveness of any process is crucial for continually improving it. Moreover, user awareness and training play important roles. Microsoft mandates all its engineers to undergo security training and to attend a refresher course once a year, in order to keep abreast of the changing trends in security attacks. Though not much information is released to the public on the specific security metrics, it is apparent that Microsoft measures the results from security metrics it collects from its projects. Some metrics include training coverage, rate of discovered vulnerabilities and attack surface.

Microsoft SDL focuses on the technical aspects of developing secure software, and does not explicitly cover the business and nontechnical requirements, such as regulatory compliances and so on. Moreover, Microsoft SDL does not cover activities related to deployment and configuration, secure operation of software, secure disposal of software and so on.

When compared to older software (that has not been subject to the SDL), new software (that is developed using SDL guidance) does, in fact, display a significantly reduced rate of external discovery of security vulnerabilities. Microsoft started releasing the SDL and its experiences with it for public consumption from 2004, and this provides an excellent source of information for those who want to institutionalize security as part of their SDLC. For more information on Microsoft SDL, interested readers may refer to [8, 11].

2.2.3 Comprehensive Lightweight Application Security Process (CLASP)

CLASP was primarily authored by John Viega and Secure Software, Inc. [3], with contributions from IBM and WebMethods. This section gives an overview of the publicly-released free CLASP 1.0 documentation. CLASP is based on an activity-centric approach and is a set of security-focused processes (activities) that can be easily integrated into any SDLC. CLASP provides a well-organized and structured approach for locating security vulnerabilities and issues in the early stages of software development, where it is most cost-effective to fix them. As the documentation states, CLASP is an outcome of extensive work with different development teams and a compilation of best practices from public information.

CLASP is primarily targeted as a set of security-focused activities, without too many recommendations regarding the SDLC phases per se. This allows the CLASP activities to be SDLC method agnostic and practitioners can choose the set of activities that fits their processes and integrates at specific life cycle phases. CLASP 1.0 is made up of several parts, which include both security activities description and an extensive set of security resources and guidance.

CLASP 1.0 primarily defines a set of 24 security-focused activities that can be integrated into any SDLC process. These activities include not just technical but managerial activities, since CLASP believes that buy-in-from-the-top management is crucial for the success of such a program, and has even clearly defined a set of security activities targeted at managers. CLASP also provides suggestions regarding organization roles required for carrying out security activities. Besides the project manager, architect, requirements specifier, designer, implementer and test analyst roles, CLASP also introduces a new role called security auditor. The security auditor will be associated with the team throughout the development and will help in auditing the effectiveness and completeness of the security activities.

Table 2.1 summarizes the CLASP 1.0 security activities based on the roles that perform those activities.

One of the key aspects of CLASP is its comprehensive vulnerability 'root-cause' database. A good understanding of the root causes of security vulnerabilities is very important in order for all team members to be able to avoid them. CLASP 1.0 provides detailed explanations of more than 100 vulnerabilities, organized logically into several groups. The root-cause database provides information like the root cause of the vulnerability's existence, illustrative code samples, advice

Table 2.1 CLASP security activity role mapping.

Role	Activities
Project Manager	1. Institute security awareness program 2. Monitor security metrics 3. Manage security issue disclosure process
Requirements Specifier	1. Specify operational environment 2. Identify global security policy 3. Document security-relevant requirements 4. Detail misuse cases
Architect	1. Identify resources and trust boundaries 2. Identify user roles and resource capabilities 3. Integrate security analysis into source management process (Integrator) 4. Perform code signing (Integrator)
Designer	1. Identify attack surface 2. Apply security principles to design 3. Research and assess security posture of technology solutions 4. Annotate class designs with security properties 5. Specify database security configuration 6. Address reported security issues
Implementer	1. Implement interface contracts 2. Implement and elaborate resource policies and security technologies 3. Build operational security guide
Test Analyst	1. Identify, implement and perform security tests 2. Verify security attributes of resources
Security Auditor	1. Perform security analysis of system requirements and design (threat modeling) 2. Perform source-level security review

on avoidance of the same and so on. CLASP states that it updates this root-cause database on a regular basis, as and when new vulnerabilities are discovered.

Another important contribution from CLASP 1.0 is its implementation guidance targeted toward practitioners. Implementing all 24 security-focused processes may not be necessary for every project in an organization. CLASP 1.0 provides an implementation guide that helps project managers or process engineers with information like activity applicability, risks associated with not performing the activity, indicative implementation cost (man hours per iteration) and any other considerations, such as dependencies between processes and so on. This helps project managers decide which particular activities can/should be adapted and adopted.

CLASP also provides a lot of other very useful information for practitioners. This includes templates, checklists and guidelines to support the various activities. The supporting material includes detailed explanations of the set of security principles to be followed in the design, guidelines on security requirements and so on.

A more detailed coverage of the specific security activities of CLASP 1.0 is out of the scope of this chapter. Interested readers may refer to [3] for more details.

2.2.4   Build Security In

BSI [5] is a project of the Software Assurance program of the Strategic Initiatives Branch of the National Cyber Security Division (NCSD) [1] of the US DHS [12]. The Software Engineering Institute (SEI) [13] was engaged by the NCSD to provide support in the Process and Technology focus areas of this initiative. The SEI team and other contributors develop and collect software assurances and software security information that help to create secure systems. BSI publishes best practices, tools, guidelines, rules, principles and other resources that software developers, architects and security practitioners can use to build security into software at every phase of its development. BSI content is based on the principle that software security is fundamentally a software engineering problem and must be addressed in a systematic way throughout the SDLC.

The DHS Software Assurance Program is grounded in the National Strategy to Secure Cyberspace, which states that: 'DHS will facilitate a national public-private effort to promulgate best practices and methodologies that promote integrity, security, and reliability in software code development, including processes and procedures that diminish the possibilities of erroneous code, malicious code, or trap doors that could be introduced during development' [5].

There are several security-enhanced methodologies and processes, and there is no magic bullet to achieve security. BSI is a believer of this principle and is certainly not about recommending a specific security SDLC. BSI provides references to a very useful set of alternative approaches, best practices and design principles to secure software development and maintenance. It provides references not only to security-enhanced process improvement models and SDLC methodologies, but to the secure use of nonsecurity-enhanced methodologies.

BSI serves as a very good reference to the different security-enhanced methodologies, and provides lots of informative resources on security design principles, vulnerability root causes and so on. Interested readers may refer to [5] for more details.

2.3   A Typical Security Engineering Process

As demonstrated in many of the above security-enhanced process models and methodologies, development of secure software requires us to address security as part of every aspect of the development process, starting from requirements analysis and moving to architecture, to design, to implementation, to testing, to deployment and to subsequent maintenance. Developing secure software is not a task to be assigned to a few select individuals; instead, it is a combined responsibility of everyone involved in the process, including folk from both management and technical streams. It requires a fundamental change in the approach to addressing the security of software. Such an initiative has to be subscribed to wholeheartedly by the top management as well – their sponsorship is critical for the success of this initiative. Any framework that wants to address security assurance in a comprehensive manner has to deal with multiple tasks, which include:

- Acquiring or developing a well-tested security-enhanced process or methodology that best fits the development culture of the organization.
- Making the fundamental changes to the organizational process that help in smoothly implementing the security cross-checks and measuring the benefits in terms of metrics.
- Creating security awareness at each and every level in the organization. This may require running different training programs, tailored to meet the interests of their respective target audiences (e.g. managers, architects, developers, testers, auditors, etc.), and conducting relevant refresher courses for every stakeholder at least once a year.
- Equipping the consultants on job with the right kind of tools and guidance, and providing adequate training and support.
- Implementing processes to audit the security posture and manage the vulnerability life cycle.
- Implementing processes to track the latest happenings in the industry, in terms of best practices, recent trends in vulnerability attacks, new vulnerability remediation techniques and so on, and to update the organization's own processes/checklists/guidelines accordingly.

In this section, we cover a typical set of baseline security activities to be incorporated at different phases of SDLC. The focus here is on technical aspects, omitting details on other aspects such as training, management activities, organizational changes required and so on. Readers interested in a more comprehensive set of security activities are encouraged to refer to the process models like SSE-CMM [2] and so on.

Figure 2.3 shows a typical SDLC, with requirements, architecture and design, development (coding) and testing phases. It shows the minimal set of security

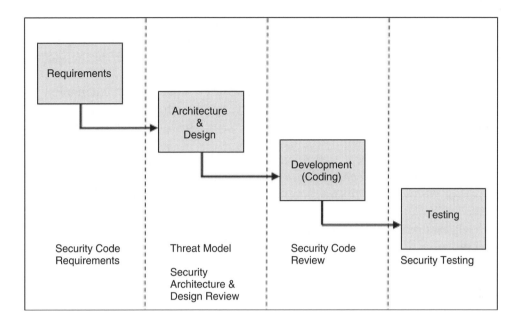

Figure 2.3 Typical security activities through SDLC.

activities recommended at each of the SDLC phases to ensure a minimum level of security assurance.

Though from the figure it looks like a waterfall model, many of these security-focused activities can be repeated at different stages of development and hence suit a spiral (iterative) model as well. The following subsection covers each of these phases in more detail.

2.3.1 Requirements Phase

Just as it is not possible to build features which are not known, security cannot be built into a product if the precise security requirements are not known up front. Requirements for security functionality in software systems are often confused with requirements for secure software. While the requirements for the security functionality of software systems are those that help implement the security policy of an organization (e.g. access control, authentication, etc.), the requirements for secure software are different and concentrate on security as a quality parameter. The requirements for secure software focuses on processes that aim at reducing software security vulnerabilities, providing software security assurance, improving dependability and so on.

Our experience shows that in most software projects, security requirements are either not captured at all or captured only partially. Most security requirement specifications are limited to aspects such as user-authentication and access-control

mechanisms only. However, there are several other dimensions to security requirements, such as nonrepudiation, exception handling, logging and so on. Common
Criteria [10] forms a very good reference for different dimensions of the security
functional requirements.

Similarly, it is very critical to capture security nonfunctional requirements,
which are quality parameters like reliability, survivability, dependability, availability and so on. It is important to capture security requirements from both these
dimensions in order to build secure software.

Security requirements specification defines the structured way in which different dimensions of the security requirements of the target application are captured
as part of the overall software requirements specification phase. Security is often
viewed as an afterthought, but experience shows that security is better addressed
through capturing security requirements up front during the requirements definition phase, and then later designing, building and testing for security. Security
requirements form the critical input for every security decision throughout the
rest of the life cycle.

2.3.2 Architecture and Design Phase

The architecture and design phase is where the architects and designers define
the overall architecture of the application in terms of the major systems and their
components, interfaces, inter-component communication mechanisms and so on.
It is this phase in which the architects design for nonfunctional requirements like
performance, flexibility and so on. To architect security into applications, it is
very important for the architects to be aware of different security threats and
their root causes. Architects have to think like hackers in order to identify the
possible security threats and plan for their mitigation. It is generally not feasible
to address every possible security threat. Architects need to make a call on how
much of the security functionality must be actually implemented to keep the risk
at a manageable level. During the architecture and design phase the architect
typically dons the hat of a security expert and uses proven security patterns and
design principles to develop the security functionality.

During this phase, the security architect typically builds threat models to analyze the security threats, uses proven security patterns and design principles, and
reviews the architecture and design from a security point of view, to ensure the
required level of security assurance. The combined wisdom gathered over many
years says that a lot of security vulnerabilities stem from architecture and design
defects. It is much more economical and efficient to identify the security vulnerabilities early in the design phase where they are introduced, rather than identifying
them during testing or, in the worst case, in a live system.

Threat modeling and security architecture and design review are considered
two important software security assurance best practices in the architecture and
design phase. Other artifacts like detailed guidelines on security design principles

and security design patterns have proven very effective in enforcing consistent security architecture.

The threat model technique helps in the systematic identification of threats, attacks and vulnerabilities that could affect the target application, so that countermeasures can be built to guard against them. There are many possible security threats and vulnerabilities, but not all of them may be applicable to the target application. Similarly, because of limited resources (time, money) or the current state of technology, not every security threat and vulnerability can be addressed in an application. As a result, every application has to live with some risk that is left after implementing all security measures. This is referred to as residual risk. A threat model helps in identifying all threats, and also in prioritizing them based on the risk exposure, so that the limited number of available resources can be utilized in an optimal manner to achieve the required level of security.

Threat modeling is typically done either as part of requirements analysis or in the early stages of the architecture and design phase. This is most effective if done in an iterative manner, where the threat model is revisited and updated after every critical SDLC milestone. Threat modeling can be used to shape the target application's design, to meet the organization's security objectives and to reduce risk. A documented threat model forms the key for all the security activities in SDLC. Different security process models use different names for this activity; 'threat model' is the name used by Microsoft SDL.

Many software development shops carry out a security architecture and design review as part of the design phase; however, such reviews typically lack security focus. It is very easy to miss security architecture and design defects unless a review is carried out with a security focus, and by eyes trained to do so. Security defects that go undetected at this phase prove very costly later, when they are caught during testing or in production. An architecture and design review that is done with a security focus using security checklists and best practices makes it easy to catch most subtle security defects early in the design stage itself, just where they are introduced.

Security architecture and design review is typically carried out by the architects, using the security guidelines/checklists/best practices available, and is done with the objective of identifying the architecture and design defects that could potentially lead to security vulnerabilities. During the review, the architects also typically look for the right usage of security patterns and secure design principles.

2.3.3   Development (Coding) Phase

Many security vulnerabilities can be primarily attributed to insecure programming practices followed by developers while coding. Security awareness is often minimal amongst the developer community. The consultants on job are not trained in security and not many schools even teach secure coding techniques as part of their curriculum. Given that even experienced developers find it difficult to identify secure coding bugs without the necessary security awareness, that fresh developers find it even more so isn't surprising. A vast majority of security vulnerabilities, like buffer overflows, injection vulnerabilities and so on, can be easily addressed by simply following good secure coding practices and principles.

A security code review is when a code review is conducted with the objective of finding security bugs. This is a critical activity in the SDLC, where most of the security vulnerabilities can be caught accurately even before the testing phase begins.

While code reviews form a regular part of any software development lifecycle, many of the current industry processes don't have the rigor that is required to identify security bugs. Security bugs are often difficult to find and special guidance can be required to uncover them. Though code reviews are looked at as daunting tasks involving substantial manual effort, they are the ideal place to capture a lot of security vulnerabilities. A common experience of organizations that do security code reviews is that they form the most effective way to identify security bugs early and thereby help in reducing the overall cost.

Security code review typically follows a checklist-based approach, where experienced developers review the code to find potential security defects. The focus is on uncovering insecure coding practices, and also on ensuring that the code is in compliance with enterprise secure coding guidelines. A recommended best practice is to de-link the security code reviews from the other regular code reviews so that the focus on security is not diluted. Moreover, repeated code reviews (at different levels, such as with peers and with senior developers) are seen to be very effective. Code that is scrutinized by multiple pairs of eyes is more likely to have security bugs detected early on.

2.3.4 Testing Phase

Traditional testing processes concentrate only on functional aspects, at the expense of security-focused test cases. This phase provides the last opportunity for the organization to identify bugs before the application goes live and hackers get their first crack at doing so.

Security testing is the process that defines how to perform a black box-based security analysis. The main objectives of software security testing include verifying that the software does not exhibit any exploitable features and that the features are implemented in a secure manner. While security testing can also be used to verify that the software implements the security functionality well, its primary objective is to verify that the business functionality is implemented in a secure manner. A typical security testing exercise follows a risk-based approach, where the testers use fault injection techniques to verify that the software hasn't made any incorrect assumptions that might compromise the overall security of the system.

The processes described above (security requirements, threat model, security architecture review, security code review, security testing) form a minimal set of security activities for software security assurance, but by no means a comprehensive one. For an enterprise to integrate security as an integral part of its development culture a lot of other aspects like user education, training, continuous research and so on need to be addressed. Also, any process that cannot be measured for its effectiveness is rendered ineffective; hence the processes related to software security assurance measurement are also necessary. These include security metrics identification, collection and so on. Different organizations follow different software development methodologies, such as waterfall, iterative, agile methods and so on. This section has tried to provide an indicative list of baseline security activities that can be adapted and used with any development methodology that the organization may choose to go with. Similarly, different security-enhanced process methods provide different techniques to carry out these activities; the interested reader is referred to [6] for a detailed coverage of the state of the art in software security assurance and security in the software development lifecycle.

2.4 Important Security Engineering Guidelines and Resources

2.4.1 Security Requirements

Security requirements come from different sources, which include business people, policies, regulations and so on. Capturing good requirements is a difficult task, and capturing good security requirements a much harder one. A good security requirements specification needs the requirements specifier to be aware of the threat environment of the target application, regulatory compliance requirements to be met, security policies of the organization, security classification of the information dealt with by the applications, business needs, user constraints, knowledge of evolving security vulnerabilities and so on. Some of the common reasons why security requirements engineering is difficult include:

- Stakeholders are not good at specifying security nonfunctional requirements and assume they are in place, developed by the software vendor.
- Security is considered as an afterthought, as it is perceived to constrain the performance and system usability.
- Specifying different ways in which bad things could impact the system if they happen needs creativity, experience and an abuser's mindset. It is very difficult to specify what a system should not do or protect from.
- The available requirements engineering process focuses more on functional requirements, not on security.

These reasons necessitate multiple specialized roles to be included as part of the security requirements capturing process, including: stakeholders, users, managers, developers, testers, security engineers, risk analysts, security auditors and so on.

There are two types of security requirement. The first one is about security functional requirements like authentication, access control and so on. The second is about nonfunctional requirements, which are often constraints on the system, expressed as negative requirements.

Security mechanisms are often confused with security requirements and as a result we commonly see security mechanisms specification in place of security requirements. There are several dimensions in which the security requirements of an information system can be analyzed. Common criteria provide a good reference to the security functional requirements of applications. Similarly, Federal Information Processing Standards (FIPS) 200 [14] describes minimum security requirements for Federal information and information systems. Unlike common criteria, FIPS 200 covers security areas including operational, technical and management aspects and is a very useful set of references for security requirements. Other good sets of references for security requirements engineering include the columns from Firesmith [15]. Some of the high-level technical security functional requirements areas include:

Identification – how a system recognizes the entities (humans/systems) interacting with it.

Authentication – how a system validates the identity of entities.

Authorization – what privileges are to be provided to an entity interacting with a system?

Nonrepudiation – how a system prevents entities from repudiating their interactions with it.

Integrity – how a system protects information from intentional or unintentional modification/tampering.

Auditing – a facility by which a system allows auditors to see the status of the security controls in place.

Privacy – how a system prevents the unauthorized disclosure of sensitive information.

Availability – how a system protects itself from intentional disruptions to service so that it is available to users when they need it.

Security nonfunctional requirements are often constraints on the software system, expressed as negative requirements (e.g. the system must not be vulnerable to injection-based attacks) which are not testable until they are mapped as positive functional requirements (e.g. the system must have a proper input validation strategy to restrict, deny and sanitize all forms of input to prevent injection attacks).

Security requirements engineering is an active research area and several innovative methods are proposed in the literature as to how security requirements can be best elicited, analyzed and captured. These include the use of misuse or abuse cases, attack trees, threat models, risk-based approaches, security use cases and so on. While some of them are very useful in identifying the security threats, others are useful in capturing the security requirements of the system. Please see [6] for more details.

2.4.2 Architecture and Design

Architecture and design helps map the requirements to the components of the application. Identification and prioritization of threats (as part of the threat model process) to the application in the target environment and a critical security review of the application architecture can increase the likelihood of identifying possible vulnerabilities early in the SDLC, when they are introduced. This greatly reduces vulnerabilities due to weak architecture and design choices.

A standardized set of security design principles and security architecture patterns will help in implementing proven security best practices. Similarly, during this phase, a critical review of the security implications of integrating any third-party products will help in detecting possible vulnerabilities in the application, and hence provide an opportunity to address them right at beginning.

Security design principles are well covered in several of the books and articles published on this subject. While the following subsections mention a very few well-known security principles, there are many others, such as economy of mechanism, reluctance to trust, securing the weakest link, complete mediation and so on. A detailed coverage of all these security design principles is out of the scope of this book; interested readers may refer to [3, 5, 6].

2.4.2.1 Principle of Least Privilege

This principle advocates the assignment of only a minimal required set of rights (privileges) [3, 5] to a subject that requests access to a resource. This helps in reducing the intentional or unintentional damage that can be caused to the resource in case of an attack. For example, for an application that just needs to read the data from database tables, it is sufficient for the application to be granted read-only access to the requisite tables, rather than read/write. Implementing this principle often requires a very granular definition of privileges and permissions.

2.4.2.2 Principle of Separation of Privilege

This principle [3, 5] states that a system should not allow access to resources based upon a single condition. Before allowing access, the system should ensure multiple conditions are met. This means the system should be abstracted into different components, where each component requires different conditions to be met. This principle indirectly ensures accountability, as the system requires multiple conditions (assuming each is owned by a different subject) to be met. This principle also localizes the damage arising from a vulnerability to a small portion of the system, while other components that require different conditions can still function.

2.4.2.3 Defence in Depth

The key idea behind this principle is to have multiple redundant defence mechanisms [3, 5]spread across different layers of the system architecture. This helps in enhanced security as the attackers have to circumvent multiple defence mechanisms to break into the system. Having varied security mechanisms across different layers can further strengthen the protection, as it means the attacker has to break the controls not only at multiple layers, but of different mechanisms altogether. Defence in depth promotes redundant security controls and avoids single points of failure.

2.4.2.4 Fail Securely

This principle [3, 5] requires the system to go into a default secure state when it fails. On failure, the system should immediately move itself into a secure condition and should not leak any sensitive information that an attacker could potentially misuse. On failure, attackers should not be given access to unauthorized resources.

2.4.3 Secure Coding

Code review from a security perspective provides the best gate by which most security bugs can be detected before testing. However, code review is a daunting job that requires huge manual efforts, given the size of the code base of current enterprise applications. Secure coding guidelines and a checklist prepared based on them form the most critical key factor for successful security code review. Recently, security as part of code review has been gaining popularity and plenty of secure coding guidelines have been getting published. Microsoft has released a very comprehensive and useful set of secure coding guidelines for its popular .NET languages [11]. Similarly, CERT [16] started an initiative to compile secure coding practices for C++ and Java environments. C++ guidelines have been published and a preliminary version of the secure Java coding guidelines is available. Sun [17] has also published a minimum set of security coding guidelines for the Java environment, though it has not been updated in recent years. Furthermore, there are a couple of popular books [18] on secure coding guidelines which deserve a place on every developer's bookshelf.

While a manual security code review is the best method from a security assurance point of view, with the increased complexity and the huge code base of the current enterprise systems, it is very laborious and often not a practical idea for many organizations. Security code reviews can be well supported through the use of automated static code analyzer tools. A tool can never be a replacement for the human brain; however, security code review tools help in improving productivity by making non-security experts be productive in security code reviews. Also, the tool usage helps non-experts to learn the security bugs over time and eventually to become experts. Today, more mature static security code analyzers [19] that accurately detect security defects with minimum false positives are available commercially. However, they are far from human perfection and are not suitable to detect all possible security vulnerabilities. A general recommended best practice is to use a combination of both manual and tool-based techniques. Use the automated tool-based scanning first to identify critical sections that need a manual code review. [19] provides references to plenty of security code analyzers that are available both commercially and from the open-source community.

2.4.4 Security Testing

Testing the security of a system is the last gate through which to detect any possible security vulnerabilities before production. There are several sorts of security testing technique in use today, of which black-box security testing (or vulnerability assessment) is the most commonly used. The black-box security testing technique provides an attacker's view of the application to the testers, and the idea is to be innovative in how the system can be compromised on its security objectives. The most typical practice is to feed insecure payloads to the application and analyze the response to detect any possible bugs. Usually, testers maintain a list of known security vulnerabilities and the insecure payloads for each. To get control over the data being transacted with the application, the testers use a wide variety of tools, which include proxies, special plug-ins to the browsers that give more control in editing the user input inline and so on. There is also a new breed of automated vulnerability assessment tool that is available both commercially and from open-source communities, which helps in automating the security testing [19].

While the security engineering support tools (secure static code analyzers, vulnerability scanners [20]) help in improving the productivity of the testers in detecting security bugs, they all suffer from false positives (something the tools identify as a security problem, but which in reality is not a bug). The security tools market is evolving and more mature products have started coming up. [20] provides references to some of the popular commercial and open-source web application vulnerability scanning tools.

2.5 Conclusion

Defective software causes several billions of US dollars' worth of financial losses every year across the globe. Fixing a security bug only after a vulnerability has been reported and/or exploited is the least desirable situation and is often very expensive in terms of financial loss and brand loss. Organizations are increasingly realizing that it is much more cost-effective if the security vulnerabilities can be prevented from getting in the software in the first place. Finding and fixing security vulnerabilities in the development phase is the best possible way of addressing security. Security engineering [23] as a subject is essentially about the activities involved in engineering a software application that is secure and reliable and has safeguards built in against security vulnerabilities. In this chapter, different security-aware software development lifecycle process models and processes have been discussed, including SSE-CMM, Microsoft SDL and CLASP. A core minimal set of security processes to be followed at different stages of the software development lifecycle and some important guidelines on these processes have been described.

References

[1] National Cyber Security Division, Department of Homeland Security (DHS), http://www.dhs.gov/xabout/structure/editorial_0839.shtm.

[2] Systems Security Engineering Capability Maturity Model (SSE-CMM), Model Description Document, version 3.0, July 15, 2003, http://www.sse-cmm.org/index.html.

[3] Dan, G.*The CLASP Application Security Process*, Secure Software Inc.

[4] Davis, N. *Secure Software Development Lifecycle Processes:* a Technology Scouting Report, Noopur Davis, CMU Technical Note CMU/SEI-2005-TN-024, Software Engineering Institute, 2005.

[5] Build Security In at https://buildsecurityin.us-cert.gov/.

[6] Goertzel, K.M. (2007) Software Security Assurance – State of the art report (SOAR), July 31, 2007, Joint endeavor by IATAC with DACS. 392 pages. A report by Information Assurance Technology Analysis Center (IATAC), an Information Analysis Center within the Defense Technical Information Center (DTIC), DoD, USA.

[7] Security in the Software Lifecycle, Making Software Development Processes – and Software Produced by Them – More Secure, DRAFT Version 1.2 - August 2006, Department of Homeland Security, accessible https://buildsecurityin.us-cert.gov/daisy/bsi/87-BSI.html.

[8] Lipner, S. and Howard, M. *The Trustworthy Computing Security Development Lifecycle*, Microsoft Corporation, 2005.

[9] *Software Security – Building Security*, Gary McGraw. Addison Wesley Professional, January 23, 2006, ISBN-10: 0-321-35670-5

[10] Common Criteria for Information Technology Security Evaluation, Part 2: Security functional components, September 2006, Version 3.1, Revision 1, accessible http://www.commoncriteriaportal.org/files/ccfiles/CCPART2V3.1R2.pdf.

[11] Patterns and Practices Security Engineering Index, at http://msdn.microsoft.com/hi-in/library/ms998404(en-us).aspx

[12] Department of Homeland Security, http://www.dhs.gov/index.shtm.

[13] Software Engineering Institute, www.sei.cmu.edu/.

[14] Minimum Security Requirements for Federal Information and Information Systems, FIPS 200, NIST, accessible http://csrc.nist.gov/publications/PubsFIPS.html.

[15] Firesmith, D.G. (2003) Engineering security requirements. *Journal of Object Technology*, **2** (1):53–68.

[16] CERT's secure coding initiative, http://www.cert.org/secure-coding/.

[17] Secure coding guidelines for the Java Programming Language, version 2.0, http://java.sun. com/security/seccodeguide.html.

[18] Howard M. and LBlanc, D. *Writing Secure Code'*, 2nd edn, Microsoft Press, Paperback, 2nd edition, Published December 2002.

[19] Source Code Static Analyzers list at http://samate.nist.gov/index.php/Source_Code_Security_ Analyzers.

[20] Web Application Vulnerability Scanners list at http://samate.nist.gov/index.php/Web_ Application_Vulnerability_Scanners.

[21] Security Considerations in the Information System Development Lifecycle, NIST Special publication 800-64, Rev1, accessible http://csrc.nist.gov/publications/nistpubs/800-64-Rev2/SP800-64-Revision2.pdf.

[22] The Economic Impacts of Inadequate Infrastructure for Software Testing, http://www.nist.gov/ director/prog-ofc/report02-3.pdf

[23] Open Web Application Security Project (OWASP), http://www.owasp.org.

3

Common Security Issues and Technologies

3.1 Security Issues

3.1.1 Authentication

The process of ensuring that the individual is indeed the person who he/she claims to be is crucial in any computing scenario. This process is called authentication. Authentication is a fundamental step, required before allowing any person/entity to carry out an operation on a computer or to access any part of a system. The person, technically termed as the principal, usually interacts with the system by providing a secret or a piece of information which the principal alone knows or is able to generate.

3.1.2 Authorization

A common security need is to provide different levels of access (e.g. deny/permit) to different parts of or operations in a computing system. This need is termed authorization. The type of access is dictated by the identity of the person/entity needing the access, and the kind of operation or the system part needing to be accessed. The access control can be enforced in many forms, including by mandatory access control, role-based access control and discretionary access control. We shall explain each of these in the following subsections.

3.1.2.1 Discretionary Access Control

Usually different principals need to be provided different levels of access for the same component; this is facilitated by capturing different privileges and

Distributed Systems Security   A. Belapurkar, A. Chakrabarti, S. Padmanabhuni, H. Ponnapalli, N. Varadarajan and S. Sundarrajan © 2009 John Wiley & Sons, Ltd

permissions for different principals. A common manifestation of such permissions is sometimes extended to groups of individuals too. Such principal-focused access control is termed discretionary access control. This is enforced by attaching access-control lists (ACLs) with each principal. ACLs are typically maintained in databases or in file systems.

3.1.2.2   Role-Based Access Control

Typical enterprise users usually perform a specific role at any point of time. Further, the access to any operation or any system is usually dependent upon the role of the principal. For example, any principal in the role of an administrator needs a different level of access to regular users. Such access-control mechanisms, where access is based on the role of the principal, are termed role-based access control (RBAC). RBAC models require maintenance of the list of roles, and mappings from a role to a group of users.

3.1.2.3   Mandatory Access Control

In many access-control scenarios it is necessary to provide access to resources based on certain discrete levels associated with the principal. The level is also associated with resources. If the principal's level is higher than the level of the resource, access is granted. This kind of access control is termed mandatory access control (MAC). This is simpler to enforce than RBAC, as we do not need to keep detailed ACLs. Only a hierarchy of access control levels need be maintained.

3.1.3   Data Integrity

In the case of online systems, it is necessary to make sure that during transmission from one location to another, a piece of data arrives at the target destination without having been tampered with. It needs to be ensured that the arrived data is correct, valid, sound and in line with the sender's expectations. This requirement, usually termed as data integrity, is key in any online transaction. It could be prevented from being achieved due to multiple factors, including deliberate tampering in transit, transmission errors, viruses or even problems caused by natural disasters. Common techniques to achieve data integrity include message authentication codes (MAC) and digital signatures.

3.1.4   Confidentiality

The most important requirement in the case of business transactions, namely confidentiality, refers to the need to restrict access to any information to authorized persons only, and to prevent others from having access to that information. Breach of confidentiality must be avoided to ensure that unauthorized personnel do not access data. For example, it is important that we do not say our passwords out loudwhen we speak on a phone, or transmit passwords via any written or online medium.

Specifically, it is important that, even though it might be theoretically possible for unauthorized persons to get access to data, this is made as difficult as possible. The aim of encryption techniques is precisely to achieve this.

3.1.5 Availability

Any piece of information must be available to authorized users when they need it. This requirement of availability states that systems be architected so that information is not denied to authorized users for any reason. Availability could sometimes be thwarted due to physical or communication factors like disk crash or improper communication channels. However, from an attack perspective, the typical hindrances to availability of a service occur in the form of the popular attacks, termed as denial of service (DOS) attacks.

3.1.5.1 Denial of Service Attacks

DOS attacks are specific attacks on a network or a computational resource, designed to make that network/resource unavailable. These DOS attacks are typically carried out via exploitation of vulnerabilities in a piece of software or in a network protocol. Hence, protocol and software designers need to constantly innovate and release newer versions to plug the gaps so that DOS attacks are minimized. In fact, in later chapters, we shall study detailed versions of the different DOS attacks at host, infrastructure, application and service layers. Typical DOS attacks aim at making a resource unavailable by flooding the system with multiple requests so that no more slots are available for legitimate consumers. Sometimes, this can even be caused by software, termed as malware, which is specifically designed to cause loss of availability, and cannot be trusted.

3.1.6 Trust

'It takes years of hard work to establish trust but only a few seconds of madness to destroy it.'

The maxim *'Trust then verify'* should be applied to distributed systems.

Trust has always been one of the most significant influences on customer confidence in services, systems, products and brands. Research has shown that trust in brands or companies often has a direct correlation to customer loyalty and retention. Trust begets customer loyalty. Consumers who have a high level of trust in their bank are more likely to perform a wider variety of more complicated online banking tasks, such as automated bill payment or applying for new products or services.

Surveys have revealed that consumers with a high level of trust in their primary bank are loyal – they aren't seeking services from other institutions, with a majority not having visited another bank's Web site. However, the studies also clearly reveal most consumers with high trust in their primary bank say they would cease all online services with their current bank in the event of a single privacy breach. That could translate into the potential loss of millions of customers, making even a single breach a very costly problem for banks. Although gaining and maintaining consumer trust is challenging, it must be a priority. Building consumer trust in the Web channel will impact customer acquisition and retention rates.

People often tend to be transparent in a known '*circle of trust*'. It is well documented that people share their passwords, ATM PINs, e-mails and so on among their perceived 'circle of trust'. However it has always been difficult for systems to be designed for this 'perceived trust'.

3.1.6.1 Definition of Trust

As per the ITU-T X.509 [1], Section 3.3.54, trust is defined as follows:

> 'Generally an entity can be said to "trust" a second entity when the first entity makes the assumption that the second entity will behave exactly as the first entity expects.'

The following principles are generally impertinent for trust modeling in security architecture.

(1) Trust is a key attribute of security architecture.
(2) The security architecture should ensure a trust level that is in the comfort zone of most of the users of the system.
(3) Trust is also a mirror of security as a whole. It is a trade-off between balancing of risk and efforts needed to mitigate that risk. There must always be a symmetric view of trust.
(4) Trust is the enabling of confidence that something will or will not occur in a predictable and defined manner. This could extend to other quality-of-service attributes like availability, reliability and so on as well.

3.1.7 Privacy

Privacy, a broader issue than confidentiality, is about the provision for any person, or any piece of data, to keep information about themselves from others, revealing selectively. Privacy issues are sweeping the information-security landscape, as individuals demand accountability from corporations, organizations and others that handle their data. In today's world of outsourcing and off-shoring, customers are very wary about privacy of their personal data and enterprises are investing a lot to ensure that their customers continue to trust them. Consumer surveys have frequently shown that the Number 1 reported answer is to limit the sharing of personal information with third parties [4].

Recently a lot of targeted e-mail phishing scams have been reported. It is not surprising that a survey points [4] to identity theft as the biggest customer concern in the event of a breach or violation of personal information. The recent controversies around loss of personal information like social security details and credit card information have also been linked to antisocial activities. For example, in 2004, an unscrupulous employee at AOL sold approximately 92 million private customer e-mail addresses to a spammer marketing an offshore gambling Web site [2]. In response to such high-profile exploits, the collection and management of private data is becoming increasingly regulated. This has led to the establishment of new laws to ensure information security. As an information-security professional, it's important that you have a basic understanding of data privacy laws and know the legal requirements your organization faces if it enters a line of business regulated by these laws.

The Health Insurance Portability and Accountability Act (HIPAA) contains a substantial Privacy Rule that affects organizations which process medical records on individual citizens. HIPAA's 'covered entities' include health-care providers, health-care clearing houses and health-care plans.

The HIPAA Privacy Rule requires covered entities to inform patients about their privacy rights, train employees in the handling of private information, adopt and implement appropriate privacy practices and provide appropriate security for patient records.

The most recent addition to privacy law in the United States is the Gramm-Leach-Bliley Act of 1999 (GLBA [3]). Aimed at financial institutions, this law contains a number of specific actions that regulate how covered organizations may handle private financial information, the safeguards they must put in place to protect that information and prohibitions against their gaining such information under false pretenses.

Data privacy is a complex and rapidly-changing field. The legal landscape surrounding it is fluid and subject to new legislation and interpretation by government agencies and the courts. Organizations creating, managing and using the Internet will often need to state their privacy policies and require that incoming requests make claims about the senders' adherence to these policies. Basic privacy issues will be addressed by providing privacy statements within the service policy. More sophisticated scenarios, involving delegation and authorization, will be covered in specifications specific to those scenarios.

A customer could state a set of 'privacy preferences', which could set the limits of acceptability and the underlying contexts. The customer could also decide the parameters allowing applications dealing with their personal information to act on their behalf.

3.1.8   Identity Management

Identity management is a process in which every person or resource is provided with unique identifier credentials, which are used to identify that entity uniquely. Identity management is used to control access to any system/resource via the associated user rights and restrictions of the established identity.

Examples in our daily life of identity management systems could include our citizenship cards, driving licenses or passports. When a citizen enters a country displaying their passport, by virtue of the citizenship rights associated with that person's identity, they can enter that country. The passport also gives them access to the resources of the country, and to perform certain operations, for example voting.

Identity management systems are important in today's enterprise context in the management of large numbers of users/employees. These systems automate a large number of tasks, like password synchronization, creation of user identities, deletion of user identities, password resetting and overall management of the identity life cycle of users. A key benefit of such systems is that when new applications are provisioned in an enterprise, they can leverage the identity management system data, without the need for specialized data for the new application.

Some specialized identity management system functionalities include a single sign-on usability imperative, wherein a user need not log in multiple times when invoking multiple applications and can reuse the logged-in status of a previous application in the same session. Single sign-on is a key imperative for enterprise usage of identity management solutions.

3.2   Common Security Techniques

3.2.1   Encryption

An important technique that is employed for the purpose of achieving confidentiality is encryption. The notion of encryption is based on the idea of transforming a piece of text (plaintext) into an encoded text (ciphertext), such that it is extremely difficult to guess the original plaintext even if someone gets access to the ciphertext. Two types of encryption mechanism exist: symmetric and asymmetric (also known as public key cryptography).

Symmetric encryption uses a single secret key to transform plaintext into an unreadable message. In order to retransform the unreadable message back into plaintext, people need to exchange the secret key between them. The secrecy of key exchange has to be ensured if we want to achieve confidentiality.

Asymmetric encryption, on the other hand, uses a pair of keys, one the private key and the other the public key. The private key is a secret known only to the owner, and the public key is known to all. A message encrypted with the owner's public key can only be decrypted with the owner's private key and in this way confidentiality is ensured. A message encrypted with the owner's private key can only be decrypted with the owner's public key, thereby providing a mechanism for preserving integrity.

3.2.2 Digital Signatures and Message Authentication Codes

A key requirement in enterprise security is that of integrity. Message authentication codes (MAC) and digital signatures have emerged as the forerunner technologies to ensure message integrity. The process of digitally signing a document begins with the taking of a mathematical summary (called a *hash code* or *digest*). This digest is a uniquely-identifying digital fingerprint of the document. If even a single bit of the check changes, the hash code will dramatically change. The next step in creating a digital signature is to sign the hash code with a key; in the case of MAC it is the symmetric shared key, while in a digital signature it is the private key. This signed hash code is then appended to the document. This becomes the MAC/digital signature of the document.

The recipient of the message can verify the hash code of the sender, using the shared key in a MAC or the sender's public key in a digital signature. At the same time, a new hash code can be created from the received document and compared with the original signed hash code. If the hash codes match then the recipient has verified that the document has not been altered. In a digital signature, the recipient also knows that only a particular sender could have sent the document because only they have the private key that signed the original hash code.

3.2.3 Authentication Mechanisms

Various techniques have evolved over the years for persons (principals) to authenticate with systems. Some of the common techniques that have emerged as popular mechanisms include:

(1) password-based mechanisms
(2) certificate-based mechanisms
(3) biometrics-based mechanisms
(4) smart cards-based mechanisms.

3.2.3.1 Passwords-Based Mechanisms

A popular form of authentication, the password refers to a secret value uniquely associated with the principal. When the principal enters the password to gain access to a system, the system verifies the equality of the entered password with the stored password, and accordingly authenticates. While user authentication based on passwords is a common technique, the path between the principal and the authentication system needs to be secure to ensure that the shared secret, that is, the password, does not get revealed. Some of the best practices to avoid common issues like password stealing/hacking include changing of passwords at regular intervals and inclusion of a combination of numeric, alphabet and special characters as part of a password.

3.2.3.2 Certificate-Based Mechanisms

A popular form of authentication on the Internet. A principal is uniquely identified by a digital certificate in many online systems. A digital certificate contains the unique public key associated with the principal, information on an authority which certifies the public key, the dates of validity of the certificate and a signature generated by the issuer.

3.2.3.3 Biometrics-Based Mechanisms

A biometrics-based authentication mechanism is based on a physical aspect of a principal that can be used to uniquely identify that person. In most cases, it is either the retinal characteristics or the fingerprints of the person. The basic premise here is that the physical characteristic is unique to the principal. However, the issue in such a system is that if a hacker somehow gets access to the physical part (finger or eye), access to the system cannot be stopped, because generating a new finger or new eye is difficult.

3.2.3.4 Smart Cards-Based Mechanisms

A popular form of authentication used by banks. A principal is uniquely identified by the information encoded physically on a card with an embedded microprocessor. These cards are now very often used in ATMs and online kiosks. The risk in a smart cards-based mechanism is the necessity to store the physical card securely in a safe place.

3.2.4 Public Key Infrastructure (PKI)

The infrastructure to support the ecosystem of managing public key systems is paramount to making a scalable secure ecosystem of broad-based public authentication systems. The infrastructure to manage public keys is based on the notion of digital certificates for validating users. While the need for a digital certificate in ensuring authentication is not doubted, the real issue is the problem of evolving a scalable mechanism for managing the proliferation of public digital certificates. The notion of a certification authority (CA) comes in handy when it comes to managing the life cycle of digital certificates, right from issue to revocation.

3.2.4.1 Components of PKI

A public key infrastructure needs diverse infrastructural elements to cover the diverse services which manage the life cycle of the public keys. Some of the key elements of the PKI infrastructure include the CA, registration authorities (RA) and repositories.

Certification authority

The role of a CA is to sign the digital certificates and make available their public keys to all users. A common mode of accessing public keys of popular CAs is via Web browsers, which typically contain the preconfigured certificates of many common trusted CAs.

CAs issue certificates with different valid periods. Each CA periodically maintains a certification revocation list (CRL), which manages certificates that have been revoked before the expiry of their validity. Clients can enquire into the validity of such URLs by consulting the CRLs.

Registration authority

A registration authority in a PKI system is responsible for verifying that a certificate requester has a valid reason to have a digital certificate, as well as other attributes like physical address and so on. It then follows an established procedure for issuing the certificate, like sending a one-time PIN to the physical mail address of the requester.

Repositories

A repository is an online, publicly-accessible system for storing and retrieving digital certificates and other information relevant to certificates, such as CRLs. In effect, a certificate is published by putting it in a repository. Typically repositories use some form of directories, such as 'lightweight' directory access protocol (LDAP) directories or X.500 directories.

3.2.4.2 Services of PKI

The key services in the context of security for public systems, typically government systems, provided by PKI cover the whole range of tasks for managing digital certificates. The important services among them include:

(1) *Issuing of certificates:* A key function of PKI is to combine physical and other verification mechanisms to ascertain the identities of certificate requestors, and issue the certificates.
(2) *Revoking of certificates:* To manage the proliferation of digital certificates, by using CRLs the infrastructure should be able to revoke public key certificates when required.

(3) *Governance:* The PKI infrastructure, including the constituent RAs, CAs, and so on, should establish the policies that govern the issuance and revocation of certificates.
(4) *Archival:* The infrastructure should archive the information needed to validate digital certificates at any date.

3.2.5 Models of Trust

Various models have evolved over the years for establishing trust in distributed systems. Some key models are explained below.

3.2.5.1 Implicit Trust Model

In this trust model, there are no explicit mechanisms for validation of credentials. An example is an e-mail originating from a sender. It is assumed in most cases by the recipient, particularly in a known domain, to have been actually sent by the apparent sender. This exhibits an implicit trust model. In today's digitally secure e-mail world there is a lot of spam and fraud, and hence this model does not fit in completely in a business realm. It also depends on the criticality of the underlying data. This trust model is also known as an assumptive trust model. This model is unobtrusive and inexpensive but is also prone to higher risks and susceptible to frauds. The trust relationships that are formed are complex and governed by a lot of human and technology factors.

3.2.5.2 Explicit Trust Model

This model of trust is used when we perform an entity confirmation in isolation without dependence on any other entity or preexisting credentials. This is the most commonly used trust model in the industry. The biggest advantage of an explicit trust model is that the authentication of the credentials is done using self-reliant mechanisms without any delegation. This leads to a higher degree of trust, with every entity associated with the trust mechanism. This trust model is required to reduce the liability of organizations and also to comply with regulatory policies.

The most common examples of explicit trust models are password-based authentication or even the PKI architectures. The password authentication is controlled by the host system and may have multiple underlying authentication algorithms. In a typical PKI architecture, the CA initiates all trust relationships. The CA is the common trust entity that performs all original entity authentications and the generation of credentials that are bound to specific entities. Though this model provides a high level of trust, it requires more effort and is traditionally expensive. However, this model is a prerequisite for financial transactions like payment gateways and e-commerce.

3.2.5.3   Intermediary Trust Model

This model of trust is used when trust or 'proof of trust' is transmitted through intermediaries. It is commonly used in peer-to-peer and distributed systems.

An example would be the following:

You are throwing a party and you invite friend B through a direct trust model. Friend B in turn comes with his friend C. If you trust friend C, it would mean you have switched to an intermediary trust model. The intermediary trust model can be complex and extremely contextual. There can 'n' levels associated with an intermediary trust model. It also involves selective associative trust models based on certain policies for certain contexts.

Enterprises today use an intermediary trust model within their own boundaries, but use it selectively with business partners and the extended enterprise.

3.2.6   Firewalls

As we saw in the section on the issue of availability, it is important that systems be rescued from diverse threats to availability, including DOS attacks, malware attacks and other such factors leading to loss of availability.

A firewall is a specialized software/hardware which inspects any network access made to a private network. It is deployed to regulate the flow of traffic from a public network to a private network. In this context, the regulation of the traffic happens via certain rules, which are configured in the firewall.

Firewalls can be of multiple forms and multiple techniques. The most common form of firewall includes those at the network level, which work by packet filtering, These packet-filtering firewalls inspect every incoming packet and, depending upon the fields, reject or accept packets to go through. These decisions are made based on filtering rules.

Likewise, proxy firewalls are meant to control access to external sites by providing a layer between the private network and the external network, in some cases helping optimize the performance of the network and Internet access by prefetching.

Yet another category of firewall includes application-level firewalls, which are aware of the needs of a consuming application. Depending upon the need of an internal application, we can configure more application-specific filter rules; further, these application firewalls work at the application layer of the network stack.

3.3   Conclusion

In this chapter, we have summarized some key security issues and concerns of interest to distributed system building. We shall elucidate detailed layer-specific issues and concerns in later chapters. This chapter is just meant to provide a heads-up on the basic relevant concepts, with more details in the later chapters.

Later, we shall explore some typical solutions and solution mechanisms in distributed systems to enable realization of architectures addressing the key issues and concerns outlined in this chapter.

References

[1] ITU-T Recommendation X.509, ISO/IEC 9594-8: Information Technology – Open Systems Interconnection – The Directory: Public-Key and Attribute Certificate Frameworks, accessible http://www.iso.org/iso/iso_catalogue/catalogue_tc/catalogue_detail.htm?csnumber=43793.
[2] Oates J. (2005) AOL Man Pleads Guilty to Selling 92m Email Addies, *The Register*, http://www.theregister.co.uk/2005/02/07/aol_email_theft/.
[3] Graham-Leach-Biley Act, details at http://www.sec.state.ma.us/sct/sctgbla/gblaidx.htm.
[4] (2005) Privacy Trust Survey for Online Banking, details at http://findarticles.com/p/articles/ mi_m0EIN/is_2005_April_5/ai_n13506021 .

4

Host-Level Threats
and Vulnerabilities

4.1 Background

Whether in a traditional n-tier system or a parallel computing infrastructure such as grid or cluster, threats to host (server as well as client) in a distributed system are numerous. These threats arise due to either mobile codes (both trusted and untrusted) that are downloaded and executed or vulnerabilities in pieces of trusted software installed on the host that could be exploited. Figure 4.1 depicts a broad classification of host threats.

4.1.1 Transient Code Vulnerabilities

'Transient code' is any binary, object code or script that is mobile and executes on a remote host. A remote host is one which executes the mobile code either intentionally or accidentally. An end user on the host could inadvertently download and execute a mobile code that has malicious intent, which could compromise the security of the host and pose serious challenges to all other hosts on the network. Attacks through Trojan horses, spyware and eavesdropping are common in the distributed world, particularly through content or applications published on the Internet. The need to deliver rich and intelligent content to users has forced the networked applications such as Web browsers, e-mail clients and so on to allow scripts to be executed, for example Java script, or an Applet or ActiveX object. A person with malicious intent can use this scripting ability to take control of a host or cause permanent damage to its content.

In cooperative computing environments such as grid or cluster, the host trusts remote jobs to behave. The reality, however, is that there are jobs that spy on the host and share confidential information about it, jobs that destroy or corrupt

Distributed Systems Security A. Belapurkar, A. Chakrabarti, S. Padmanabhuni, H. Ponnapalli, N. Varadarajan and S. Sundarrajan © 2009 John Wiley & Sons, Ltd

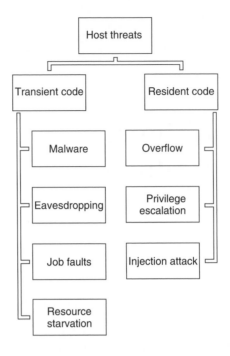

Figure 4.1 Classification of host-level threats.

content, and less seriously, jobs that hog host resources and don't relinquish them to other, native jobs.

4.1.2 Resident Code Vulnerabilities

'Resident code' is a piece of trusted software installed on the host with the knowledge of the user. Resident code, however trusted, may contain some vulnerabilities. Historically, most attacks have happened by exploiting these vulnerabilities. See Figure 4.2.

4.2 Malware

Spybot, Sasser, Trojan.Spy.Win32.Logger and Stash are perhaps familiar to most people, and security experts recognize them as the top malwares to have affected hosts worldwide in recent times. Some malwares are Trojan horses, some are spyware and some are worms or viruses, but all can pose a serious security threat to hosts [1]. In the rest of this section, we briefly look at each of these types of threat and at the vulnerability of hosts to them.

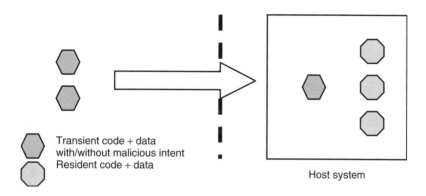

Transient code + data
with/without malicious intent
Resident code + data

Host system

Figure 4.2   Resident and transient codes.

4.2.1   Trojan Horse

'Trojan horse' [2] in the context of systems normally refers to a software façade that seems genuine or innocuous, but has the potential to be destructive. Often the victim is unaware of the fact that their system has been compromised. Unlike other malware, such as viruses or worms, Trojan horses cannot operate autonomously. Trojan horses typically reach the victim in the form of an attachment in an e-mail or data in a sharable media such as a CD-ROM, a flash drive and so on, or as a file sent through online chats/messenger or downloaded from a Web site/ftp site. In May 2000, an e-mail message originating from East Asia, widely recognized as 'Love Bug', caused havoc by flooding e-mail systems worldwide and led to severe financial losses. This was a simple e-mail, which attracted people's attention through its subject line. It contained an attachment that was in vbscript. When an unsuspecting victim opened the attachment, the script distributed itself to others in their address book.

Trojan horses are scripted with the intention to create back doors for remote access to the host under attack, disable the security scanner/firewall setup on the host and prepare the host to participate in a distributed denial-of-service or even crypto-viral attack. Attractive screen savers, catchy attachment names and games are some of the methods that hackers have used to infect hosts. In some operating systems, file extensions are hidden to make it easy for users to manage their information. Often this has been exploited to infect an unsuspecting host.

4.2.2   Spyware

Spywares [3] are programs that covertly install themselves through some back-door mechanism, including Trojan horses, and spy on a user's private/ confidential data. Once installed, spyware also monitors the user's access

behaviour, collecting information and using it to make commercial gains for its
author. SearchSeekFind, ShopAtHomeSelect and Surfairy are three well-known
spywares from recent times.

4.2.3 Worms/Viruses

Unlike Trojan horses and spyware, which largely rely on social engineering
and user action, viruses and worm [4] have the ability to spread on their own.
Viruses are generally malicious in intent and are created by their authors to
replicate themselves on all executable files and macros on a host. Viruses
then rely on the users to share infected files with users on other hosts.
Michelangelo, Brain and Jerusalem are among the popular viruses that infected
many PCs/desktops in the 1990s. As systems began to network, virus authors
found an easier means to infect hosts over networks. This led to the advent of
worms. Compared to viruses, worms adopt a more active propagation technique
and use vulnerabilities in networks and hosts to infect them. Nimda, Melissa,
Sasser and Code Red are three popular worms, which caused havoc and brought
down several hosts concurrently, by exploiting the fact that systems were
networked heavily and were running software with vulnerabilities such as buffer
overflow.

4.3 Eavesdropping

In collaborative computing platforms such as grid, business users often share
their desktops and other computing resources. Some of these users are genuinely
interested in sharing their computing resources for a common cause; others have
malicious intent, and still others inappropriately handle information that passes
through their system. There are three classes of vulnerability under this category, namely unauthorized access to confidential data, unauthorized access to
protected or privileged binaries and unauthorized tampering with computational
results (Figure 4.3). Similarly, there are jobs that users submit on the grid which
are malicious in intent. Jobs may eavesdrop into execute nodes' protected data
and share this with the job submitter. We discuss each of these vulnerabilities in
depth in this section.

4.3.1 Unauthorized Access to Confidential Data – by Users

Consider a lengthy computation that involves analysis of critical business performance data. Often the timeliness of the report is very crucial. Its availability within
a stipulated period could be the difference between staying competitive and losing
business. This problem makes a compelling case for a distributed solution. However, there are security challenges in implementing this as a distributed solution
in a cooperative resource-sharing platform [5].

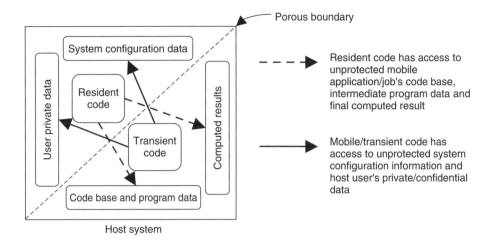

Figure 4.3   Eavesdropping vulnerability.

Typically, in such computing environments there are schedulers, which identify a node for job execution based on machine load and uptime. Once a node is identified, the scheduler provisions both the executable binaries/script and the relevant data. As the grid system does not assume any privileges on a remote host, all the files related to this job are hosted in a folder (such as tmp) to which every logged-on user on the system has access.

If the scheduled job runs on the host even for less than a minute, users on the host with malicious intent have adequate time to copy sensitive data to an alternative location. They can then examine, analyze, interpret and misuse the data offline, and employ brute force or similar techniques against encrypted staged data.

Another serious issue may arise if staged data contains some access privilege information such as a user ID and password or other authentication keys. This is required if the scheduled job has to impersonate the user or use alternative credentials to access some external systems. Such access credentials, if they become available to a user with malicious intent, can have severe security repercussions which are difficult to trace or identify.

In summary, grid or similar jobs, typically provisioned on remote hosts for execution along with their data, become vulnerable to attacks from users who either own the execution host or have access to it.

In a more conventional distributed system such as a Web application, user interactions are through a Web browser. The limitation of Web browsers and the HTML language in providing a rich interface [6] has led to several innovations, such as Applets, scripting and ActiveX controls. All these mechanisms require the Web browser to download and execute mobile alien code on the host. This can pose serious threats to the system, ranging from eavesdropping on private information, through corrupting system data, to taking complete control of the system.

Netscape introduced cookies to overcome challenges in HTTP protocol (particularly related to HTTP being stateless). Web sites use cookies to store information on a user's machine, so that some information about the user's interaction with the site is available the next time they access it. Cookies by themselves are harmless and do not pose any security threat. However, the data available in cookies can provide insights into the user's browsing habits, which malicious users can use for commercial or other gains. Cookies which store access control information are even more dangerous. They can be accessed by a sniffer on the network, and reused to impersonate the user and steal their identity.

4.3.2   Unauthorized Access to Protected or Privileged Binaries – by Users

In this subsection we turn our attention to issues that may arise out of inappropriate use of job binaries.

First, a binary may actually implement some proprietary algorithm or trade secret, and a malicious user gaining access to such a piece of code or executable could cause an organization great losses. For example, take a rate calculator used by an insurance firm. Apart from the service and other product features, insurance rate/premium often provide competitive advantage to an insurance company and any theft of this information could lead to severe losses.

Second, some binaries contain certain access privilege information, such as a user ID/password, hardcoded in them. Merely copying such a binary and reusing it would provide a malicious user with elevated privileges to a system. For example, applications that connect to a database often store the database access credentials in the code or some libraries. Malicious users can use these to gain access to the database system. This may not only give unprecedented access to sensitive information, but makes data manipulation/tamper a definite possibility.

Finally, malicious users may simply re-execute the binary without having any knowledge of the data on it. If the binary happens to alter some external system, it can perhaps make the system inconsistent and affect business. In this case, the intent of the user is not to gain information or access to a privileged system, but to cause inconvenience or disturbance.

To summarize, access to grid job binaries by malicious users can negatively affect a system and is clearly a security vulnerability.

4.3.3   Unauthorized Tampering with Computational Results

Next, we look at security vulnerability in a collaborative compute platform related to action by a malicious user. In this case, the malicious user alters the results of a computation with the intent to corrupt or mislead the system [7]. It is rather difficult to initiate such attacks, and they are largely dependent on the type of application, the output it generates and the duration of the run.

Typically, such attacks are possible only if the application's output structure is well understood and it runs for a certain minimum duration. It is difficult to tamper with the results of very short-lived jobs that run for less than a second. Similarly, jobs that post their computation results directly to a centralized system without storing them in intermediate files are difficult to tamper with. Jobs whose source code is available, and which run for several minutes to hours and use intermediate files to store results, are often easy targets for such attacks. Reports show that data returned by nodes on the SETI@Home project are at times not reliable due to tampering.

4.3.4 Unauthorized Access to Private Data – by Jobs

Earlier in this section, we discussed potential security vulnerabilities which could be exploited by malicious users. In this subsection we focus on the issues arising from malicious jobs.

Though grid schedulers tend to deprivilege grid jobs and let them run with minimal privileges on the host, most hosts leave a lot of sensitive data accessible to anyone with basic privileges. Potentially, the grid jobs which run on a remote host could exploit this vulnerability to read and expose private and sensitive information stored there. For instance, if not properly secured, password shadow/cache files on the host could be used to get access to it. A host of system configuration and other files might also be vulnerable to such attacks. See Figure 4.4.

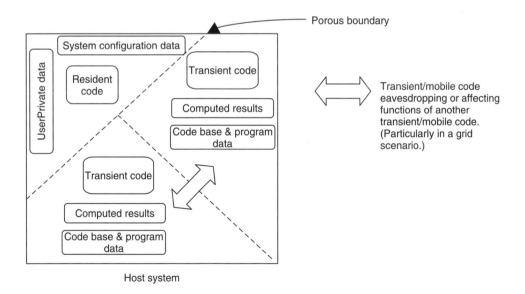

Figure 4.4 Transient code eavesdropping/affecting other code.

While hosts are vulnerable to attacks from mobile jobs that execute on them, equally vulnerable are other mobile jobs that execute at the same time. By default, most grid schedulers allow more than one job to run on the host concurrently. The number of jobs typically allowed on a host is equal to the number of CPU slots available. This creates the possibility of sensitive information from an innocuous job falling into the hands of a malicious one which happens to execute concurrently on the same host. In this case, the host facilitates the attack without being aware of it.

4.4   Job Faults

Often with enterprise applications built on traditional three-tier architecture or packaged applications such as SAP, Oracle and so on, the application binaries or scripts are well known. However, in the case of ad hoc distributed computing platforms such as grids, which may be used heavily by research departments or engineering departments to solve computationally-intense or simply long-running batch applications, the complete inventory of possible applications/binaries is not available a priori. Engineers, researchers and other users write applications that solve specific problems they are working on. These applications do not normally undergo quality tests or certification. When scheduled to run over the grid, they can potentially cause faults (though in this case, the faults are generated unintentionally), which can bring down the entire host, along with all the applications running on it.

In other cases, the job owners may have malicious intentions and script jobs targeting a particular host to inject a fault that will corrupt the host or simply cause reboot/shutdown.

This does not imply that the host alone is vulnerable to mobile jobs that run on it. It is equally possible for the host applications and users to be harmful to the grid applications and to inject faults that cause the grid applications to either fail or behave inappropriately.

4.5   Resource Starvation

In the literal sense, resource starvation [8] may not be termed entirely as a security vulnerability. However, it is important to consider it, simply because it is a contentious issue when it comes to building a distributed shared computing platform such as grid out of nondedicated compute resources such as production servers and desktop machines. The primary objectives of production servers and desktop machines are different. Grid is just one of the applications of these compute resources and in many cases it does not even feature in the top three in terms of importance. In this context, any job/application that negatively affects the primary objective of the resource (desktop/production servers) is more than an irritant and is undesirable.

Grid applications tend to be resource intensive. They are either CPU-bound or memory-bound and are written in such a way that there are hardly any synchronization or other waits. Once initiated, they saturate the hosts, on which they run completely, leaving very little behind for other applications to use. For instance, a computational fluid dynamics application works with a large data set consuming most of the physical memory on the host and, depending on the problem size, can run for long periods, while soaking up all available CPU slots on the host.

In today's enterprise, most compute resources, particularly the mid-range UNIX and lower-end x86 servers, are very poorly utilized, and 24-hour utilization in many cases hovers below the 10% threshold. The challenge really is to use this spare capacity without affecting the applications/jobs already running on these hosts. The easiest way to use this idle capacity is to augment this to an existing grid or cluster. The resource starvation issue makes it difficult to guarantee quality of service to both existing applications on the host (native applications) and the grid applications. In the case of production servers which are used as grid nodes, the response times or the throughput of the native host applications are affected because of the computationally-intense nature of the grid jobs. Similarly, in the case of desktop nodes, the interactive nature of the host may be in jeopardy. In either case, resource starvation on the grid node affects the smooth operation of the node's primary function.

Job starvation occurs when a long-running job is ahead of the job queue, blocking out several reasonably smaller jobs. This kind of job starvation is more to do with job and queue management in a grid or cluster platform and less to do with hosts, and is not in the scope of this book.

4.6 Overflow

Buffer overflow [9] (see Figure 4.5) vulnerabilities constitute nearly half of all vulnerabilities reported by CERT. In 2001, a worm known as Code Red exploited over a quarter of a million systems that were on the IIS Web server. This worm exploited the buffer overflow vulnerability that existed in the IIS implementation. Buffer overflow is a vulnerability on the host that can lead to memory access exceptions, predominantly left in the system due to programming oversights or errors.

Buffers are used by software programs to store data. They typically have an upper and a lower bound. Any location outside these bounds does not belong to the program. Either accessing or storing data in locations outside the bounds leads to program faults or unexpected behavior. Numerous host vulnerabilities in the past

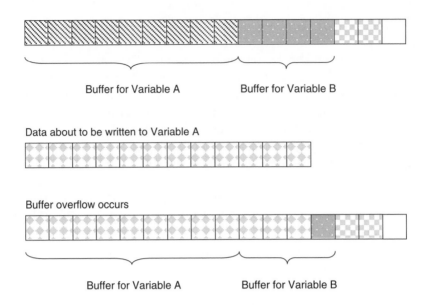

Figure 4.5   Buffer overflow.

have arisen due to buffer overflow issues. Among the overflow vulnerabilities, the most important are stack-based buffer overflow and heap-based buffer overflow. These are discussed further below.

4.6.1 Stack-Based Buffer Overflow

Programs use a stack segment to store local variables and function return information. A malicious user may manipulate inputs or bytes transferred over the network to the program to corrupt the stack, thereby altering the behavior of the program to suit their needs.

The first type of stack-based buffer overflow [10] threat attempts to overwrite adjacent locations in the stack, modifying variables either intentionally or accidentally. This causes erratic or incorrect program behavior, sometimes causing the application to fail abruptly. Exploiting such vulnerabilities, it is possible for a user with malicious intent to bring down a critical service or application on a host. Each day security experts are identifying more and more stack-based buffer overflow vulnerabilities and threats on the host.

The second type of stack-based buffer overflow threat modifies a function pointer or return address to execute arbitrarily different code. Programs store the function argument and the return location in stack frames. Program control transfers to a location specified in the altered value when the buffer overflows. Malicious user may carefully place the block of code they would like to execute and smash the stack to jump to this program location.

Lower address ←	Stack segment	→ Higher address	
Buffer	Function code segment base pointer	Function code segment instruction pointer	Function arguments

Figure 4.6 A typical stack before overflow attack.

Lower address ←	Stack segment	→ Higher address
NOP + Exploit code	New exploit code's base pointer	New exploit code's instruction pointer

Figure 4.7 Stack after overflow attack.

NOP-sleds and jumps to registers are common techniques used by malicious hackers to exploit any stack buffer overflow vulnerability that may exist in a program. See Figures 4.6 and 4.7.

4.6.2 Heap-Based Buffer Overflow

Heap-based buffer overflows are different from the stack-based buffer overflows in the sense that the data contained in the heap are predominantly program data allocated dynamically during the lifetime of the program's execution and do not contain function pointers/return location. Though stack-based buffer overflows tend to outnumber heap-based buffer overflows, there are still a few vulnerabilities on the host that arise due to heap overflows. For instance, the heap overflow vulnerability in the Microsoft Telnet server is well known.

4.7 Privilege Escalation

Privilege escalation [11] is an issue related to an unauthorized user elevating their authorization level on a host in order to perform tasks that they are not otherwise allowed to perform. In most cases, the elevated privilege desired by malicious users is that of root or the user equivalent of root. By elevating privilege to that of root, a malicious user can take absolute control of the host. This can have severe repercussions, including the host being part of distributed denial-of-service or other similar attacks, which are difficult to trace. There are two possible types of privilege escalation, namely horizontal privilege escalation and vertical privilege escalation.

In the case of horizontal privilege escalation, a malicious user tries to assume the identity of a peer in the system. The objective of the malicious user is not to take complete control of the machine, rather they are trying to access another user's private data or conceal their own identity and present themselves to others as the compromised user.

Vertical privilege escalation occurs when a malicious user assumes the identity of a system administrator (root in Linux or LocalSystem account in Windows). This allows the malicious user to access all information stored on the host and perform tasks which are otherwise not permissible.

Buffer overflows are often the easiest way to achieve privilege escalation. Hackers have exploited buffer overflows more than anything else to gain root privileges on a host, as is apparent from the CERT vulnerability reports [12].

4.8 Injection Attacks

Injection attacks are attempts by malicious hackers to exploit an application vulnerability that does not handle user inputs securely. Injection attacks generally use an executable code/script to intrude into a host system. There are several ways to inject malicious code through user inputs. Shell/PHP injection [13] and SQL injection [14] will be explored briefly in this section.

4.8.1 Shell/PHP Injection

In this type of injection attack, an execute shell command or script is fed as input to a field, with the knowledge that the text will form part of an executable command string.

Consider the following 'C' code:

```
...
sprintf(buffer, "grep {-}i %s x.txt", argv[1]);
    /* form cmd string */
system(buffer); /* execute command string */
```

Running the above executable passing an argument formatted as "search-string; init 6; echo " would cause the system to reboot, provided the script is running with root privilege. As is evident from the example, it is possible to run arbitrarily any command on the host. The entire host will be under the attacker's control if the vulnerable application is running with root privileges.

PHP/ASP injection commonly refers to such a code injection vulnerability on applications built over a server-side scripting engine such as PHP or ASP.

4.8.2 SQL Injection

Similar to the shell injection, an SQL injection attack exploits vulnerabilities in applications involving databases. In this type of attack, a malicious user uses an insecure application input field to inject a harmful query string or retrieve unauthorized data. The injection attack is best illustrated with an example. Consider an input field that accepts an authentication code and an ID for access control, and internally uses a query similar to this:

```
"SELECT * FROM USERS WHERE ID = ' " + id + ' " AND
    ACCESSCODE = ' " + AccessCode + " ' "
```

The system accepts the ID and AccessCode as inputs from the users. The system does not perform any tests to validate the inputs. A malicious hacker can now use this to stage an SQL injection attack.

First, we look at an input a hacker can use to bypass the access control applied through this SQL. Using a value such as "'password' OR 'A' = 'A'" for Access-Code, the entire 'where' clause is made redundant. Next we look at an input which can cause some serious damage to the application data. Using a value "'password'; DELETE FROM USERS WHERE 'A' = 'A'", the entire access control table can be wiped out. It is evident from the examples that SQL injection attacks can be serious threats to application security.

Certain database servers allow execution of external operating system shell commands through special stored procedures. A malicious hacker can use this to cause serious irrecoverable damage to the database server host.

4.9   Conclusion

We have walked through several host-level threats and vulnerabilities that can cripple the host and other interconnected systems. We broadly classify applications running on the host as either resident code or transient code depending on whether the user/administrator on the host trusts the piece of code. Security vulnerabilities and threats on hosts are entirely dependent on the code that is running on them. Security vulnerabilities introduced due to transient code are of particular importance given that new and emerging paradigms such as grid/cluster and more mainstream platforms based on intranet/Internet rely on transient code executing on remote hosts. Hosts are equally vulnerable, if not more so, to trusted resident code, mainly due to engineering faults in them.

We have classified all the host vulnerabilities discussed thus far based on the impact of the vulnerability and on the period in which this assumes significance (Table 4.1). We classify a vulnerability to be of high impact (can cause serious damage to monetary or reputation, or human loss), medium impact (some damage to business or reputation) or low impact (marginal loss commercially, very insignificant). Similarly, we classify period of impact as immediate/near-term, medium-term (three-to-five years) or long-term (beyond five years).

If we look at host vulnerabilities exposed by transient code, malwares tend to be authored with malicious intent and are normally destructive in nature. The

Table 4.1 Summary of the host-level threats.

Attack	Impact	Period	Remarks
Transient code			
Malware	High	Immediate	Attacks by viruses, worms and Trojans on a host could cause serious irrecoverable damage to the host and the business the host supports. This requires immediate attention
Eavesdropping	High	Immediate	For mainstream intranet/Internet application, eavesdropping can lead to security and privacy breach. For newer platforms such as grid, the issue is not of immediate significance
Job faults	Medium	Medium	These are particularly important in the context of shared computing environments. This threat is not in the critical path
Resource starvation	Low	Long	This again is particularly focused on shared computing platforms such as grid. The threat is not immediately significant
Resident code			
Overflow	High	Immediate	Any resident code vulnerability has the potential to cause serious damage to the host as well as the applications running on it and in a highly networked environment, it has a tendency to spread fast. These security vulnerabilities require immediate attention
Privilege escalation	High	Immediate	
Injection	High	Immediate	

impact of such attacks is high and they need immediate attention. User privacy/confidentiality is assuming more importance every day and any breach to this has social and legal implications. In this context, any issue relating to eavesdropping has to be classified as a high-impact issue which requires immediate solution. Resource starvation and prevention against faults are not issues related to mainstream applications and they may assume significance in the future as grid and other shared computing/scavenging platforms become pervasive.

In the case of resident code, the security vulnerabilities need to be treated rather seriously. There is a general assumption that transient code is rogue, but the same cannot be said about resident code. It is resident and voluntarily installed on the host by the user/administrator under the premise that it is from trusted, genuine sources. All vulnerabilities under this category are to be treated as high-impact as well as immediately relevant.

References

[1] (2008) The Crimeware Landscape: Malware, Phishing, Identity Theft and Beyond, http://www.antiphishing.org/reports/APWG_CrimewareReport.pdf, accessed on June 12th 2008.

[2] Schneier, B. (1999) Inside Risks: the Trojan Horse Race. *Communications of the ACM*, **42** 9, 128.

[3] Sipior, J.C., Ward, B.T. and Roselli, G.R. (2005) A United States Perspective on the Ethical and Legal Issues of Spyware. Proceedings of the 7th International Conference on Electronic Commerce.

[4] Kurzban, S. (1989) Viruses and worms - what can they do? *ACM SIG SAC Review*, **7** 1, 16–32.

[5] Waldburger, M. and Stiller B. (2007) Regulatory Issues for Mobile Grid Computing in Europe. 18th European Regional ITS Conference (ITS 2007), September 2007, Istanbul, pp. 1–10.

[6] (2008) Ajax Security Dangers, http://www.spydynamics.com/assets/documents/AJAXdangers.pdf, accessed on June 12th 2008.

[7] Golle, P. and Mironov, I. (2001) Uncheatable distributed computations. In David Naccache, editor, *Topics in Cryptology – CT-RSA 2001*, volume 2020 of Lecture Notes in Computer Science, pages 425–440. Springer, April 2001.

[8] Chakrabarti, A., Grid Computing Security, pp. 154–156, Springer, 2007, ISBN: 3540444920.

[9] (2008) Buffer Overflow Demystified, http://www.enderunix.org/docs/eng/bof-eng.txt, accessed on June 12th 2008.

[10] (2008) Smashing The Stack For Fun And Profit, http://insecure.org/stf/smashstack.html, accessed on June 12th 2008.

[11] Chess, B. and West, J., Secure Programming with Static Analysis, pp. 421–456, ISBN: 0-321-42477-8, Addison Wesley, Published: June 2007.

[12] (2008) CERT Vulnerability Statistics, http://www.cert.org/stats/fullstats.html, accessed on June 12th 2008.

[13] Younan, Y., Joosen, W. and Piessens, F. (2004) *Code Injection in C and C++: A Survey of Vulnerabilities and Countermeasures*. Technical Report CW386, Departement Computerwetenschappen, Katholieke Universiteit Leuven, July 2004.

[14] Breidenbach, B. (2002) Guarding your Web Site against SQL Injection Attacks, ASP Today.

5

Infrastructure-Level Threats and Vulnerabilities[1]

5.1 Introduction

In the real world when the word 'infrastructure' is mentioned it refers to the highways, roads, bridges, flyovers, power grids, transmission towers and other structures which support the movement, communication, power and other essential elements of human society. There are similar elements which support the basic functioning of IT systems, like the networking infrastructure, the middleware and the storage infrastructure. Securing the IT infrastructure is being identified as critical by different government agencies, as attacks may have serious consequences on the security and the economic vitality of a society [1]. As Richard Clarke, former US homeland security advisor for combating cyber terrorism, puts it (CNN News, October 9, 2001), 'Our very way of life depends on the secure and safe operations of critical systems that depend on cyberspace.' In this chapter, we acknowledge the importance of IT infrastructure security and focus on the threats to and vulnerabilities of the same.

5.2 Network-Level Threats and Vulnerabilities

The most critical component of the IT infrastructure is the networking infrastructure. It forms the core of any system and is similar to the roads and highways in real life. However secure we make our cars, if the roads and highways are fraught with security dangers, the whole system suffers. The networking infrastructure has seen a huge growth over the last few years, especially with the advent of wireless

---

[1] Contents in this chapter reproduced with permission from Grid Computing Security, Chakrabarti, Anirban, 2007, XIV, 332 p. 87 illus., Hardcover, ISBN: 978-3-540-44492-3

Distributed Systems Security A. Belapurkar, A. Chakrabarti, S. Padmanabhuni, H. Ponnapalli, N. Varadarajan and S. Sundarrajan © 2009 John Wiley & Sons, Ltd

technologies. But research and development efforts have been mostly in the areas of performance and scalability, with security taking a back seat. The importance of securing the network has grown rapidly in recent years due to the series of attacks which shut down some of the world's most high-profile Web sites, like Yahoo! and Amazon. Several examples of such attacks can be found in CERT reports. Securing the networking infrastructure is clearly the need of the hour and different components of the networking infrastructure, like the routers, servers, wireless devices and so on, need to be protected for a sustained IT security.

5.2.1 Denial-of-Service Attacks

One of the most dangerous network-level threats is the denial-of-service (DoS) attack. These attacks have a simple objective, to deny service to the service consumers. This is generally achieved by overwhelming the networking infrastructure with huge numbers of data packets. In DoS attacks, the packets are routed correctly but the destination and the network become the targets of the attackers. DoS attacks are very easy to generate and are very difficult to detect, and hence they are attractive weapons for hackers. In a typical DoS attack, the attacker node spoofs its IP address and uses multiple intermediate nodes to overwhelm other nodes with traffic. DoS attacks are typically used to take important servers out of action for a few hours, resulting in DoS for all users. They can also be used to disrupt the services of the intermediate routers. Generally, DoS attacks can be categorized into two main types: (i) ordinary and (ii) distributed. In an ordinary network-based DoS attack, an attacker uses a tool to send packets to the target system. These packets are designed to disable or overwhelm the target system, often forcing a reboot. Often, the source address of these packets is spoofed, making it difficult to locate the real source of the attack. In the distributed denial-of-service (DDoS) attack, there might still be a single attacker, but the effect of the attack is greatly multiplied by the use of attack servers known as 'agents'. To get an idea of the scope of these attacks, over 5000 systems were used at different times in a concerted attack on a single server at the University of Minnesota. The attack not only disabled that server but denied access to a very large university network [2].

Experts are studying DoS attacks to identify any trends that can be inferred from attack patterns. Unfortunately, no such trends have emerged. However, the experts have unanimously declared that DoS attacks are the most potent of all infrastructure attacks. A Computer Security Institute (CSI) survey [3] shows that 30% of all attacks on the Internet are of the DoS type. Given the amount of impact these can generate, the CSI result is extremely frightening. Though the survey was done a few years ago, evidence suggests that the percentage of DoS attacks has increased rather than decreased. Figure 5.1 shows the different categories of attacker responsible for attacking the Internet infrastructure, based on the same survey. It shows

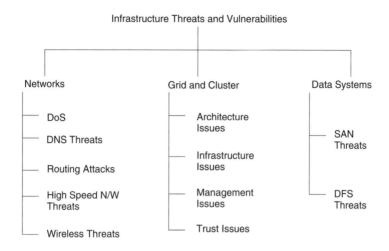

Figure 5.1 Taxonomy of infrastructure threats and vulnerabilities.

that despite the popular image of cyber terrorism and corporate-based cyber warfare, most attackers are independent hackers or disgruntled employees doing it for the fun or due to animosity against their employer.

Though there have been indications of the importance of DoS attacks, the actual data are mostly hidden because most companies prefer to keep attack stories from the public. One of the interesting works on the importance of DoS attacks was carried out by Moore *et al.* [4]. In the paper, the authors tried to answer the simple question, how prevalent are DoS attacks on the Internet today? The results are far-reaching and remain an important warning about the importance of tackling DoS attacks. As a means to demonstrate this, the authors described a traffic-monitoring technique called the 'backscatter analysis' for estimating the worldwide prevalence of DoS attacks. Using backscatter analysis, the authors observed 12 805 attacks on over 5000 distinct Internet hosts belonging to more than 2000 distinct organizations during a three-week period. The authors further estimated a lower bound on the intensity of such attacks – some of which are in excess of 600 000 packets per second. The paper showed the importance of DoS attacks in the context of the Internet.

As is quite evident from the above paragraph, DoS attack is becoming one of the most potent attacks carried out over the Internet. With what little data is available, the damages seem to run to millions of US dollars. Most of the attackers are amateurs rather than corporations or rogue countries engaged in cyber warfare. Now the natural question that comes to the mind is, how do these amateurs have the firepower to break the security of the biggest corporations of the world? There are two answers to this. First, the wide availability of DoS launching tools. If one searches for 'DoS attack tool', one will get over 1000 hits and lots of open freewares for launching DoS attacks. Second, the defense against this type of attack is still in its nascent stage, and lots of research is required to provide protection against DoS attacks. In the next section, we will concentrate on the different types of DoS attack.

5.2.1.1 Distributed Denial-of-Service Attacks

One of the deadliest forms of DoS attack is when the attackers are distributed in nature. Such an attack is called a DDoS attack. According to the computer incident advisory capability (CIAC), the first DDoS attacks occurred in the summer of 1999 [5]. In February 2000, one of the first major DDoS attacks was waged against yahoo.com. This attack kept Yahoo! off the Internet for about 2 hours and cost Yahoo! a significant loss in advertising revenue [6]. Another DDoS attack occurred on October 20, 2002 against the 13 root servers that provide the domain name system (DNS) service to Internet users around the world. They translate logical addresses such as www.yahoo.edu into a corresponding physical IP address, so that users can connect to Web sites through names (which are more easily remembered) rather than numbers. If all 13 servers were to go down, there would be disastrous problems for anyone accessing the World Wide Web. Although the attack only lasted for an hour and the effects were hardly noticeable to the average Internet user, it caused 7 of the 13 root servers to shut down, demonstrating the vulnerability of the Internet to DDoS attacks [7]. If unchecked, more powerful DDoS attacks could potentially cripple or disable essential Internet services in minutes.

Let us now discuss about some of the common DDoS attacks carried out by malicious agents. Most of these attacks target a particular network protocol, like the Transfer Control Protocol (TCP), User Datagram Protocol (UDP) and so on.

SYN flood attacks

Perhaps the most popular DDoS attack is the synchronize (SYN) flood attack. This type of attack targets the TCP to create service denial. The TCP protocol includes a three-way handshake between the sender and the receiver before data packets are sent. The protocol works in the following manner:

(1) The initiating system sends a SYN request. This indicates the system's intention to create a TCP session.
(2) The receiving system sends an ACK (acknowledgement) with its own SYN request. This indicates that the receiving system would like to carry on with the connection.
(3) The sending system then sends back its own ACK and communication can begin between the two systems. It has been proved that three-way handshake is an efficient and effective way to create a network connection. If the receiving system is sent a SYNX packet but does not receive an ACKY+1 to the SYNY it sends back to the sender, the receiver will resend a new ACK+SYNY after

some time has passed. The processor and memory resources at the receiving system are reserved for this TCP SYN request until a timeout occurs.

In a DDoS TCP SYN flood attack, the attacker instructs the zombies (systems previously compromised by the attacker for this purpose) to send bogus TCP SYN requests to a victim server in order to tie up the server's processor resources, and hence prevent the server from responding to legitimate requests. The TCP SYN attack exploits the three-way handshake by sending large volumes of TCP SYN packets to the victim system with spoofed source IP addresses, so the victim system responds to a nonrequesting system with the ACK+SYN. When a large volume of SYN requests is being processed by a server and none of the ACK+SYN responses are returned, the server begins to run out of processor and memory resources. Eventually, if the volume of TCP SYN attack requests is large and the requests continue over time, the victim system will run out of resources and be unable to respond to any legitimate users.

PUSH+ACK attacks

In this type of attack, the attacker again uses the properties of the TCP protocol to target victims. In the TCP protocol, packets that are sent to a destination are buffered within the TCP stack and when the stack is full, the packets get sent on to the receiving system. However, the sender can request the receiving system to unload the contents of the buffer before the buffer becomes full by sending a packet with the PUSH bit set to one. PUSH is a one-bit flag within the TCP header. The TCP stores incoming data in large blocks for passage on to the receiving system in order to minimize the processing overhead required by the receiving system each time it must unload a nonempty buffer. The PUSH+ACK attack is similar to a TCP SYN attack in that its goal is to deplete the resources of the victim system. The attacking host or the zombies and reflectors (any system that responds with an IP packet if sent one) send TCP packets with the PUSH and ACK bits set to one. These packets instruct the victim system to unload all data in the TCP buffer (regardless of whether or not the buffer is full) and send an acknowledgement when complete. If this process is repeated with multiple agents, the receiving system cannot process the large volume of incoming packets and will crash.

Smurf attacks

In a DDoS Smurf attack, the attacker sends packets to a network amplifier (a system supporting broadcast addressing), with the return address spoofed to the victim's IP address. The attacking packets are typically ICMP ECHO REQUESTs, which are packets (similar to a 'ping') that request the receiver to generate an ICMP ECHO REPLY packet. The amplifier sends the ICMP ECHO REQUEST packets to all of the systems within the broadcast address range, and each of these systems will return an ICMP ECHO REPLY to the target victim's IP address. This type of attack amplifies the original packet tens or hundreds of times.

5.2.2   DNS Attacks

The DNS is a distributed, hierarchical, global directory that translates machine/domain names to numeric IP addresses. The DNS infrastructure consists of 13 root servers at the top layer, and top-level domain (TLD) servers (.com and .net) as well as country-code TLDs (.us, .uk and so on) at the lower layers. Due to its ability to map human memorable names to numerical addresses, its distributed nature and its robustness, the DNS has evolved into a critical component of the Internet. Therefore, an attack on the DNS infrastructure has the potential to affect a large portion of the Internet.

Attacks of this type have illustrated the lack of authenticity and integrity of the data held within the DNS, as well as in the protocols that use host names as an access control mechanism.

5.2.2.1   Impact of 'Hacking'

DNS, being a critical infrastructure, is contacted by all hosts when they access servers and start connections. The impact of DNS attacks is quite widespread; effects include:

(1) *Denial-of-Service:* DoS is one of the most dangerous impacts of DNS 'hacking'. DoS can be achieved in several ways: one is to send back negative responses indicating that the DNS name does not exist. Another is to redirect the client's request to a server which does not contain the service the client is requesting. DoS attacks on DNS servers can achieve the same objective with greater effect.
(2) *Masquerading:* The attacker can use DNS attacks to redirect communication to masquerade as a trusted entity. If this is accomplished, they can intercept, analyze and/or intentionally corrupt the communications [8].
(3) *Information leakage:* DNS threats also include leakage of information concerning internal networks to an attacker. Frequently, host names can represent project names that may be of interest, revealing the operating system of the machine.
(4) *Domain hijacking:* By compromising insecure mechanisms used by customers to update their domain registration information, attackers can take over the domain registration process to hijack legitimate domains.

5.2.2.2   Types of 'Hacking'

DNS consists of a distributed database, which lends to its robustness and also leads to various types of vulnerability, which can be categorized into three main types:

(1) *Cache poisoning:* Generally, to hasten the process of query response, DNS servers store common information in a cache. If a DNS server is made to cache

bogus information, the attacker can redirect traffic intended for a legitimate site to a site under the attacker's control.

(2) *Server compromising:* Attackers can compromise a DNS server, thus giving them the ability to modify the data served to the users. These compromised servers can be used for cache 'poisoning' or for DoS attacks on some other server.

(3) *Spoofing:* In this type of attack, the attacker masquerades as a DNS server and feeds the client wrong and/or potentially malicious information. This type of attack can also redirect traffic to a site under the attacker's control, and launch a DoS attack on the unsuspecting client. In order to address DNS attacks, the IETF added security extensions to the DNS, collectively known as DNSSEC [9].

5.2.3   Routing Attacks

Routing tables are used to route packets over any network, especially the Internet. Routing protocols like distance vector, link state and path vector protocols have been designed to create routing tables through the exchange of routing packets. Routing table 'poisoning' is a type of attack on the routing protocols where the routing updates are maliciously modified, resulting in the creation of incorrect routing tables.

5.2.3.1   Impacts of Routing Table Poisoning

Routing table poisoning can have impacts like suboptimal routing, congestion, partition, overwhelmed host, looping and illegal access to data.

(1) *Suboptimal routing:* With the emergence of the Internet as a means of supporting soft real-time applications, optimality in routing assumes significant importance. Routing table poisoning attacks can result in suboptimal routing, which can affect real-time applications. Similarly, in a grid scenario this type of attack may lead to suboptimal routing, resulting in a Quality of Service (QoS) violation.

(2) *Congestion:* Routing table poisoning can lead to artificial congestion if packets are forwarded to only certain portions of the network. Artificial congestion thus created cannot be solved by traditional congestion control mechanisms.

(3) *Partition:* The poisoning attack may result in the creation of artificial partitions in the network. This can become a significant problem since hosts residing in one partition will be unable to communicate with hosts residing in another.

(4) *Overwhelmed host:* Routing table poisoning may be used as a weapon for DoS attacks. If a router sends updates that result in the concentration of packets into one or more selected servers, the servers can be taken out of service because of the huge amounts of traffic. This type of DoS attack is

more potent as the attacker is not spoofing identity, and is thus impossible to detect by standard detection techniques.

(5) *Looping:* The creation of triangle routing, caused due to packet mistreatment attacks, can also be simulated through improper update of the routing table. Loops thus formed may result in packets getting dropped and hence in lowering of the overall network throughput.

(6) *Access to data:* Attackers may gain illegal access to data through the routing table poisoning attack. This may lead to the attackers snooping packets.

5.2.3.2 Different Routing Protocols

Routing protocols can be broadly grouped into three main categories: distance vector, link state and path vector routing protocols.

(1) *Distance vector:* In this set of protocols, the nodes in the network create a vector of shortest path distances to all the other nodes in the network. This distance vector information is exchanged between the nodes. After receiving the distance vector information from its neighbors, each node calculates its own distance vector. One point to note about these protocols is that no node has the full topology information and each depends on its neighbors for creating its routing tables. It has been shown that several problems, like the count-to-infinity problem, can result from not having the full topology information. The Routing Information Protocol (RIP) [10] is an example of a distance vector protocol.

(2) *Link state:* In link state protocols, each node sends its connectivity information to all the other nodes in the network. Based on the information received from the other nodes, each node computes the shortest path tree by applying the Bellman Ford algorithm. Unlike the distance vector protocol, each node participating in the link state protocol has the full topology information. As a result, link state protocols are inherently robust. Open Shortest Path Forwarding (OSPF) [11] is an example of a link state protocol.

(3) *Path vector:* This protocol is a variation of the distance vector. In this protocol, each node sends the full shortest path information of all the nodes in the network to its neighbors. It has been shown that problems associated with standard distance vector protocols can be avoided in the path vector protocol. The Border Gateway Protocol (BGP) [12] is an example of a path vector protocol.

5.2.3.3 Routing Attacks

Routing table poisoning can be broadly categorized into (i) link and (ii) router attacks. Link attacks, unlike router attacks, are similar in the case of both link state and distance vector protocols.

(1) *Link attacks – interruption:* Routing information can be intercepted by an attacker, and the information can be stopped from propagating further.

However, interruption is not effective in practice. The reason for this is that in the current Internet scenario there is generally more than one path between any two nodes, since the average degree of each node is quite high (around 3.7). Therefore, even if an attacker stops a routing update from propagating, the victim may still be able to obtain the information from other sources.

(2) *Link attacks – modification/fabrication:* Routing information packets can be modified/fabricated by an attacker who has access to a link in the network.

(3) *Link attacks – replication:* Routing table poisoning can also take the form of replication of old messages, where a malicious attacker gets hold of routing updates and replays them later.

(4) *Router attacks – link state:* A router can be compromised, making it malicious in nature. Router attacks differ in their execution depending on the nature of the routing protocol. In the case of a link state routing protocol, a router sends information about its neighbors. Hence, a malicious router can send incorrect updates about its neighbors, or remain silent if the link state of the neighbor has actually changed. A router attack can be proactive or inactive in nature. In the case of a proactive router attack, the malicious router can add a fictitious link, delete an already existing link, or change the cost of a link proactively. In the case of an inactive router attack, the router ignores a change in link state of its neighbors.

(5) *Router attacks – distance vector:* Unlike with link state, in the case of distance vector protocols, routers can send wrong and potentially dangerous updates regarding any nodes in the network, since the nodes do not have the full network topology. In distance vector protocols, if a malicious router creates a wrong distance vector and sends it to all its neighbors, the neighbors accept the update since there is no way to validate it. As the router itself is malicious, standard techniques like digital signatures do not work.

5.2.4 Wireless Security Vulnerabilities

Network technologies are slowly moving in the wireless direction as more and more transactions take place using mobile systems. In developing countries like India and China, the potential of wireless networks is enormous and more and more of the rural population is jumping on the wireless bandwagon. However, even with the growth of wireless technologies, enterprises are slow in going fully mobile. Other than operational issues, security concerns are their primary reason. In this section, we will briefly outline the security concerns present in wireless technologies.

5.2.4.1 Traffic Analysis

One of the simplest attacks that can be employed against a wireless network is to analyze the traffic in terms of the number and size of the packets transmitted.

This attack is very difficult to detect as the attacker is in promiscuous mode and hence mostly hidden from any detection techniques. In addition to getting the information that there is a certain amount of wireless activity in the region, the attacker can learn the location of the access point in the area. Also, the attacker may be able to obtain information about the type of protocol used. For example, if TCP is used, a three-way handshake will be employed. Once knowledge of the protocol is obtained, further attacks like man-in-the-middle or session hijacking can be performed [13].

5.2.4.2 Eavesdropping

This is another technique that can be used to get information about the packets and data transmitted through the wireless channel. In this type of attack, the attacker is assumed to be passive, getting information about the data transmitting through the wireless channel. In addition to the payload, source and destination information can be obtained, which can be used for spoofing attacks. It is to be noted that encryption can be used to prevent this type of attack [13, 14]. However, gathering enough information through eavesdropping can help in breaking some of the simpler security encryption protocols like Wireless Equivalent Privacy (WEP).

5.2.4.3 Spoofing

In this type of attack, the attacker is not passive, but rather actively participates in the attack. The attacker changes the destination IP address of the packet to the IP address of a host they control. In the case of a modified packet, the authentic receiving node will request a resend of the packet and so the attack will not be apparent. Another approach is to resend the packet with the modified header. Since the receiver judges whether a packet is valid, the resend should not cause any response from the access point or access controller, which kindly decrypts the packet before sending it to the attack receiver, thus violating the confidentiality of the communication.

The attacker can inject known traffic into the network in order to decrypt future packets in the wireless network. This type of attack can be useful in detecting the session key of the communicating parties. Stricter measures of encryption like changing the session keys and using stronger security protocols are needed to prevent this attack from taking place.

5.2.4.4 Unauthorized Access

While the above attacks are directed at the users of the wireless technologies, this attack is directed at the wireless network as a whole. After gaining access to the network, the attacker can launch additional attacks or just enjoy free network use. Due to the physical properties of WLANs, the attacker will always have access to the wireless component of the network. In some wireless security architectures this will also grant the attacker access to the wired component of the network. In other architectures, the attacker must use some technique like MAC address spoofing to gain access to the wired component.

5.2.4.5   Man-in-the-Middle Attack

This is a classical attack which is applicable to the wireless domain as well. In this type of attack, the attacker acts as an interface between the two communicating parties. For example, let A and B communicate with one another. The attacker C gets into the middle such that A communicates with C thinking that it is B, and B communicates with C thinking that it is A. The attacker can sneak into the middle of the conversation by gaining access to header information and spoofing the header information to deceive the recipient. An ARP poison attack is one manifestation of man-in-the-middle attack. In this type of attack, the attacker sends a forged ARP reply message that changes the mapping of the IP address to the given MAC address. The MAC address is not changed, just the mapping. Once the cache has been modified, the attacker can act as a man-in-the-middle between any two hosts in the broadcast domain.

Man-in-the middle attacks can be simple or quite complicated depending on the security mechanisms in place. The more security mechanisms in use, the more mechanisms the attacker will have to subvert when re-establishing the connection with both the target and the access point. If authentication is in place, the attacker must defeat the authentication mechanism to establish new connections between themself and the target and themself and the access point. If encryption is in use, the attacker must also subvert the encryption to either read or modify the message contents. This type of attack can be used to launch eavesdropping, spoofing or even DoS attacks.

5.2.4.6   Session Hijacking

Session hijacking is an attack against the integrity of a session. The attacker takes an authorized and authenticated session away from its proper owner. The target knows that it no longer has access to the session but may not be aware that the session has been taken over by an attacker. The target may attribute the session loss to a normal malfunction of the WLAN. Once a valid session has been owned, the attacker may use the session for whatever purposes they want and maintain the session for an extended time. This attack occurs in real time but can continue long after the victim thinks the session is over. To successfully execute session hijacking, the attacker must accomplish two tasks. First, the attacker must masquerade as the target to the wireless network. This includes crafting the higher-level packets to maintain the session, using any persistent authentication tokens and employing any protective encryption. This requires successful eavesdropping on the target's communication to gather the necessary information. Second, the attacker must stop the target from continuing the session. The attacker normally will use a sequence of spoofed disassociate packets to keep the target out of the session [13–15].

5.2.4.7 Replay Attacks

This is also a very common technique which finds manifestation in wireless networks. In this type of attack, the attacker saves the current conversation or session, to be replayed at a later time. Even if the current conversation is encrypted, replaying the packets at a later time will confuse the recipient and create some other dangerous after-effects. Nonce or timestamps are generally used to prevent this type of attack from taking place. However, if the attacker is able to selectively modify the contents of the packets, this type of solution does not work.

5.3 Grid Computing Threats and Vulnerabilities

Grid computing [16, 17] is widely regarded as a technology of immense potential in both industry and academia. The evolution pattern of grid technologies is very similar to the growth and evolution of Internet technologies that was witnessed in the early 1990s. Similar to the Internet, the initial grid computing technologies were developed mostly in universities and research labs to solve unique research problems and to enable collaboration between researchers across the globe. Recently, the high-computing industries like finance, life sciences, energy, automobiles, rendering and so on have been showing a great amount of interest in the potential of connecting standalone and silo-based clusters into a department- and sometimes enterprise-wide grid system. Grid computing is currently in the middle of evolving standards, inheriting and customizing from those developed in the high-performance, distributed and, recently, web-services communities. Due to the lack of consistent and widely-used standards, several enterprises are concerned about the implementation of an enterprise-level grid system, though the potential of such a system is well understood. The biggest concerns are the security aspects of the grid [18]. The grid security issues can be grouped into three main categories: architecture-related issues, infrastructure-related issues and management-related issues.

5.3.1 Architecture-Related Issues

Architecture-level issues address concerns with the grid system as a whole. Issues like information security, authorization and service-level security generally destabilize the whole system, hence an architecture-level solution is needed to prevent these.

5.3.1.1 Information Security

We define information security as security related to the information exchanged between different hosts or between hosts and users. The concerns at the information-security level of the grid can be broadly described as pertaining to:

(1) *Unauthorized access:* Unauthorized access is one of the most dangerous attacks possible upon a grid infrastructure. An unauthorized and malicious user can gain access to a grid and get information which they were not supposed to get. If they are able to access the grid, they can launch more dangerous attacks in the form of DoS attacks. Therefore, grid security requirements should contain authentication mechanisms at the entry points. Different authentication mechanisms should be supported. It is possible to have different authentication mechanisms for different sites within a grid. Therefore, the security protocol should be flexible and scalable to handle all the different requirements and provide a seamless interface to the user. Also, there is a need for management and sharing of context.
(2) *Confidentiality:* Data flowing through the grid network, if not properly protected, can be read by unauthorized users. Therefore, the grid security mechanisms should protect the confidentiality of the messages and the documents that flow over the grid infrastructure. The confidentiality requirements should include point-to-point transport as well as store and forward mechanisms. Similar to the authentication mechanisms, there may be a need to define, store and share security contexts across different entities.
(3) *Integrity:* Grid security mechanisms should include message integrity, which means that any change made to the messages or documents can be identified by the receiver.
(4) *Single sign-on:* In a grid environment, there may be instances where requests have to travel through multiple security domains. Therefore, there is a need for a single sign-on facility in the grid infrastructure.
(5) *Delegation vulnerabilities:* There may be a need for services to perform actions on a user's behalf. For example, a computational job may require accessing a database many numbers of times. When dealing with delegation of authority from one entity to another, care should be taken so that the authority transferred through delegation is scoped only to the task(s) intended and a limited lifetime, to minimize misuse.

5.3.1.2 Authorization

Another important security issue is that of authorization. Like any resource-sharing system, grid systems require resource-specific and system-specific authorizations. It is particularly important for systems where the resources are shared between multiple departments or organizations, and department-wide resource usage patterns are predefined. Each department can internally have user-specific resource authorization as well. If we were to design an authorization system for such a library, the first thing we need to consider is that the system does not have too much overhead. In other words, there is a need to authorize users; however, there should not be a long queue in front of the library. Therefore, scalability is one of the primary concerns for designing such a system. Second, one has to keep in mind the effect should the system be tampered with. In that case, a user may be given more or less authorization than they deserve. Therefore, security is a very important concern. Third, it is possible that after a user has been authorized and allowed to enter, the authorities receive information that they are a thief. Therefore, there should be a mechanism to deny them access to resources once such information has become available. In other words, there should be a method for revocation of a user's authorization. Lastly, if different stakeholders in the library employ different authorization systems, is the current system interoperable between them? Some details about the issues are provided below:

(1) *Scalability issues:* Scalability is one of the most important and desirable characteristics of a grid authorization system. A system is said to be scalable if there is no perceived difference when many more entities access it. There are two aspects to grid scalability: one is based on the number of users, and the other is based on the amount of grid dynamism. The first is straightforward – the grid authorization system should perform well when the number of users increases. As for the second, grid systems have an inherent dynamism embedded into them. In a grid system, users may join or leave the grid system quite frequently. Furthermore, resources may be added to or removed from the grid infrastructure in an on-demand basis.

(2) *Security issues:* Like any other system, one has to analyze the security vulnerabilities existing in grid authorization systems. If an attacker hacks into a grid authorization system, one has to understand the effect of such a malicious activity. Two types of compromise are possible in a grid authorization system: user level and system level. In the former case, a user is compromised, allowing the attacker to use the grid as the user would. In the latter, the authorization system is taken over by the attacker.

(3) *Revocation issues:* Another important issue that needs to be considered before designing a grid authorization system is that of revocation of authorization. Consider the following scenario: a user logs into the grid system and is authorized to access its resources. After some time, it is learned that the user has been compromised. In this case, the user should be denied access to the resources.

(4) *Inter-operability Issues:* Different authorization systems may be used by different parties or virtual organizations and the important issue here is that of inter-operability of these different authorization systems.

5.3.1.3 Service-Level Security

The word 'service' or 'services' is finding wide usage is day-to-day business transactions. According to Merriam Webster, one definition of service is 'the occupation or function of serving' or 'the work performed by one that serves'. As one can observe, the definition of service is intrinsically linked with the service provider or 'one who serves'. Therefore, a service should always contain four basic components:

(1) A service provider or one who is providing the service to users.
(2) A set of service consumers who accesses the service provided by the service provider.
(3) A service infrastructure on which the service is provided.
(4) A set of service publishers which publish the type and nature of service provided.

We can extend the definition of 'service' and its components to real-life examples. Let us take the example of a banking service. Here the service provider is the bank, with the customer-service executives being the front-end to which the customers are exposed. The service consumers are the customers of the bank, and the service infrastructure includes the host of databases and other servers, the communication networks, and the buildings and other infrastructures that support the bank. Finally, the service publisher may be a Web site which describes the services provided by the bank, which may help the service consumers in making a service decision. Generally, a service is published in multiple channels. For example, there may be Web sites indicating the number of banking service providers in a district, and banks may have call centers to provide more details about the service they are providing.

Let us now get into the mind of the attackers who are hell-bent on disrupting the service offered by a service provider. Generally, attackers go by the principle of maximum effect. Among the four components mentioned in the previous subsection, compromising the service infrastructure or the service publisher will have the greatest effect. The reason is that if infrastructure is compromised, the service to a large number of customers is disrupted. Similarly, if the service publisher publishes wrongly or maliciously, the effect will be devastating. Again, the effects can be minimized if the infrastructure is protected, or if the publisher publishes through multiple channels. These tactics come under the purview of service disruption prevention mechanisms. If we take a look at different service providers, we find that enormous efforts are being made to make infrastructures secure. There are also laws and regulations to keep attackers from manipulating published information. This is true in the case of the physical world. Therefore, in the digital world too, there is a need for techniques and methods to counter such threats. Before looking at the different methods, techniques and research outputs available in the domain of service-level security, let us look at the different vulnerabilities and threats present there.

The different categories of threat present in services are QoS violation, unauthorized service access and DoS.

(1) *QoS violation:* Let us assume that there is a pizza delivery company whose unique selling point is that they deliver pizza within 30 minutes of the customer call. If the company is not able to deliver within the stipulated time, the customer gets a free pizza. If a malicious 'pizza eater' tries to stop the company from delivering on time, the 'pizza eater' gets a free pizza and the company loses a lot of goodwill. Now, translate the same problem to the digital world. A company may end up losing a lot of money if service level agreements (SLAs) are not met.

(2) *Unauthorized access:* In this type of threat, illegitimate or unauthorized users get access to the service. This problem is similar to the traditional problems of authentication and authorization. Standard authentication and authorization techniques discussed in this book can be used to solve this problem.

(3) *DoS attack:* Perhaps the most deadly of the service-level threats is denial of service to consumers. This type of attack is similar to the network DoS attacks described before.

5.3.2   Infrastructure-Related Issues

The grid infrastructure consists of the grid nodes and the communication network. The security issues related to the grid infrastructure are of paramount importance. Host-level security issues are those issues that make a host apprehensive about affiliating itself with the grid system. The main sub-issues here are: data protection and job starvation. Whenever a host is affiliated with the grid, one of the chief concerns is the protection of the already-existing data in the host. The concern stems from the fact that the host submitting the job may be untrusted or unknown to the host running the job. To the host running the job, the job may well be a virus or a worm which can destroy the system. This is called the *data protection* issue. Job starvation refers to a scenario where jobs originating locally are deprived of resources by alien jobs scheduled on the host as part of the grid system.

In the context of grid computing, network security issues assume significant importance, mainly due to the heterogeneity and high-speed requirements of many grid applications. Please refer to Section 5.2 for network-related vulnerabilities which can also apply to grid infrastructure. Two specific issues discussed here are the network issues related to grid computing and host-related issues.

5.3.2.1   Grid Network Issues

Currently most research and development activities in grid computing take place for the e-sciences community. The community is big and the research challenges are enormous. However, when grids move to the enterprises, several interesting and critical challenges will be witnessed. Some of the challenges and possible efforts have been highlighted in previous chapters. Another big challenge is integration with firewall technologies. Most of the enterprises employ firewalls and packet filtering, and efforts will need to be taken to solve the problem of easy integration with these.

Globus and firewall

Figure 5.2 shows the firewall requirements for different components of Globus. In the figure, a controllable ephemeral port describes the port which is selected by the Globus Toolkit, which is constrained by a configurable limit. On the other hand, an *ephemeral port* describes a nondeterministic port assigned by the system in the range less than 1024. The requirements of the different components are described as follows:

(1) *GSI:* GSI involves the authentication, confidentiality, integrity and secure-delegation modules of Globus. The request should originate from an ephemeral port and, similarly to ssh configuration, the server listens to port 22.

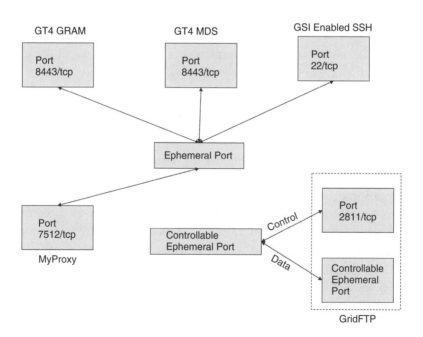

Figure 5.2   Firewall requirements for grid.

(2) *GRAM:* GRAM is the resource management module of Globus. In the GT4 GRAM, connections are initiated by the client from an ephemeral port. To initiate and control jobs, all traffic goes through a single hosting environment defined by port 8443/tcp. For GT3, this port is 8080/tcp.

(3) *MDS:* MDS is the monitoring service of Globus. Similar to GRAM, connections are initiated by the client from an ephemeral port and all traffic goes through a single hosting environment defined by port 8443/tcp. As in GRAM GT3, for MDS GT3 this port is 8080/tcp.

(4) *MyProxy:* As mentioned in Chapter 9, MyProxy is a credential storage service for X.509 credentials. MyProxy connections are authenticated and secured with GSI and are normally from ephemeral ports on the client to 7512/tcp on the server.

(5) *GridFTP:* Similarly to any FTP service, GridFTP requires two different channels: control and data channels. The control connection is established from a controllable ephemeral port on the client to the well-known static port of 2811/tcp on the server. In the case of a single data channel, the connection is established from a controllable ephemeral port on the client to a controllable ephemeral port on the server. In the case of third-party transfers (a client controlling a file transfer between two servers), this connection may be from a server to another server. In the case of multiple parallel data channels, the direction of the connection establishment is dependent on the direction of data flow – the connection will be in the same direction as the data flow.

5.3.3 Management-Related Issues

The management issues that grid administrators are worried about are credential management- (CM-) and trust management-related issues.

5.3.3.1 Credential Management

Management of *credentials* becomes very important in a grid context as there are multiple different systems, which require varied credentials to access them. CM systems store and manage the credentials for a variety of systems and users can access them according to their needs. This mandates that the CM system should provide secure transmission of credentials and secure storage of credentials, and should cater to different types of system and mechanism. Let us now look at the different characteristics that a CM system requires:

(1) *Initiation:* Every CM system should provide mechanisms so that users can obtain the initial credentials from it. The CM system should provide the required credential after authenticating the user. The authentication can be based on multiple different mechanisms, for example password-based, certificate-based and so on.

(2) *Secure storage:* As mandated by the SACRED RFC [19], the long-term credentials and the private keys should be stored in the CM systems in a

secure manner, preferably encrypted. This is a very important requirement, as compromise of the long-term credentials would have disastrous consequences.

(3) *Accessibility:* This is more related to the utility of the CM system. The CM system should be able to provide credentials when the user needs them. Proper access control mechanisms need to be provided when credentials are accessed.

(4) *Renewal:* Most credentials have a specific expiration time. The CM system should be able to handle renewal of expired credentials.

(5) *Translation:* This is important if there are multiple systems with different authentication and security mechanisms. The credentials used in one domain or realm may have to be translated into credentials in other domains, which should be handled by the CM system.

(6) *Delegation:* Delegation is really important from the grid perspective. CM systems should be able to delegate specific rights to others on the user's behalf.

(7) *Control:* Monitoring and auditing of credential usage is very important because it not only helps prevent credential compromise, it can be used for pricing if required. Therefore, CM systems should be able to monitor and audit the credentials provided to the users.

(8) *Revocation:* Finally, the CM system should provide mechanisms to revoke credentials in case of user compromise.

5.3.3.2 Trust Management

In our everyday life, we come across different types of people, situations, events and environments. Interactions between individuals depend on an implicit understanding of relationships across society. This implicit understanding between individuals is based on the confidence that the individuals experience and emanate during their relationships; the personal and professional connections bonding the individuals, societies and cultures; and a host of other factors. When we meet a new car mechanic, we feel confident to give them the responsibility of repairing a car based on our personal interactions with them, or on a hunch about their abilities, their credentials, the company they represent and so on. In other words, our decision to ask for their help depends upon the amount of *trust* we can put into the claims that they are making. Trust pervades through different strata of human society and extends beyond the human–human relationships. For example, trust can act as a means of developing relationships between electronic gadgets like computers and human beings. Trust is a complicated concept, and the ability to generate, understand and build relationships based on trust varies from individual to individual, situation to situation, society to society and environment to environment. For example, we may trust the car mechanic to repair our car. However, we may not have enough trust to allow them to repair our computer. Each individual, in general, carries some amount of prejudice based on past experiences or history, which is generally used to determine the trustworthiness of a person they are interacting with.

The Trust Management System (TMS) lifecycle is mainly composed of three different phases: trust creation phase, trust negotiation phase and trust management phase. The trust creation phase generally occurs before any trusted group is formed, and it includes mechanisms to develop trust functions and trust policies. Trust negotiation, on the other hand, happens when a new untrusted system joins the current distributed system or group. The third phase, the trust management phase, is responsible for recalculating trust values based on transaction information, distribution or exchange of trust-related information, and update and storage of trust information in a centralized or distributed manner.

Trust creation phase
This phase actually takes place before any transactions and is responsible for setting up the trust functions and the policies that will be used by the trust management system. The first step in the trust creation phase is to determine the type of trust management system: whether it will be policy-based or reputation-based. The type of TMS system is decided and the policies are defined and created. The next important step in this phase is to determine the trust function. As mentioned earlier, trust functions can be of several categories: objective or subjective, transaction-based or opinion-based, complete or localized and threshold-based or rank-based. The choice of trust function depends on the type of application the TMS is catering to.

Trust negotiation phase
The second phase in the lifecycle of the TMS is the trust negotiation phase, which begins when a new entity or node joins the system. Figure 5.3 shows a very high-level overview of a typical trust negotiation phase. The phase essentially consists of three different steps: request, policy exchange and credential exchange. At the heart of the trust negotiation lie the policies and the policy language acceptable to both parties. Several policy languages have been developed as part of the distributed trust management solutions [20]. The different steps in the trust negotiation phase are:

(1) *Request:* This step identifies the client and the type of service the client wants from the system. This step can succeed a key establishment phase, where the session key can be established by the two parties.
(2) *Policy exchange:* This step exchanges the policies between the new entity and the system. The policies can be expressed in policy languages like the PeerTrust [21]. At this step, the trust computation of the new node can also be evaluated based on the system's trust function.

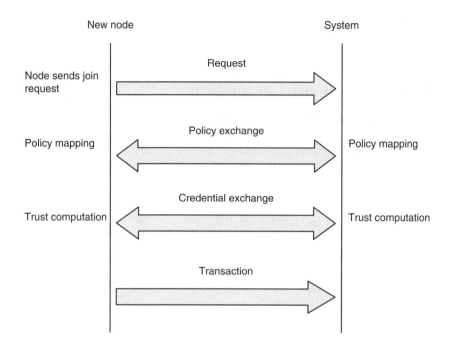

Figure 5.3   Taxonomy of infrastructure threats and vulnerabilities.

(3) *Credential exchange:* In this step, secure exchange of credentials like keys, certificates and so on takes place. Proper security measures need to be taken to ensure the secure exchange of credentials.

Trust management phase
After the trust negotiation phase comes the trust management phase, which is concerned with the general running of the distributed system. The different steps that make up this phase are:

(1) *Trust computation:* In this step, the trust value is computed based on the decided trust function.
(2) *Trust distribution:* This step includes the secure distribution of trust information to other nodes in the distributed system. Since secure distribution is a necessity, all the principles of security, viz confidentiality, authentication, integrity and nonrepudiation, need to be maintained. This step also requires keeping in mind the type of trust function in use and the number of nodes in which the information needs to be broadcast.
(3) *Trust storage:* The trust information needs to be securely stored. The credential repositories can be used for this purpose.
(4) *Trust update:* Updating the trust needs to be carried out either on an event-by-event basis or in a timely manner. Event-based trust update can

happen after a set of transactions or when the trust value or opinion crosses a threshold.

5.4   Storage Threats and Vulnerabilities

One of the most important and underestimated areas of IT infrastructure is the storage infrastructure. The importance of the storage infrastructure lies in the fact that it contains enterprise data, which is possibly the most important resource in any modern enterprise. Though extreme precautions are taken at the network, server and transport levels, the storage infrastructure must be securely protected as well. Keeping the storage unprotected is like keeping your valuables in a locker without any lock. However protected the doors, windows, elevators are, there will always be a vulnerability if the locker is kept unlocked. Similarly, whatever strong protection there may be at the different layers of an enterprise IT stack, if the storage remains unprotected, the enterprise will always be vulnerable. In this section, we will briefly look at the vulnerabilities present in centralized and distributed file systems.

5.4.1   Security in Storage Area Networks

Storage area networks (SANs) have become de facto storage infrastructures in most enterprises. SANs are high-speed data storage networks that connect different types of storage device. One of the most popular modes of storage communication is the Fibre Channel fabric. In this subsection we will touch upon the Fibre Channel protocol, before discussing the threats to SANs.

5.4.1.1   Fibre Channel Overview

Fibre Channel [22, 23] uses frames for communication purposes. Similar to the TCP/IP layer, Fibre Channel works on the principle of network layers. Each layer in the Fibre Channel network interacts with the layers below and above to transmit the frames to their destination. Most SANs use either a switched Fibre Channel topology, similar to what we use in an IP-enabled switch network, or a Fibre Channel arbitrated loop (FC-AL). In either topology, each layer performs a specific function depending on the architecture that has been deployed. The five different layers of Fibre Channel frames are as follows:

(1) *FC-0 or the physical layer:* The lowest level (FC-0) defines the physical links in the system, including the fibre, the connectors, and the optical and electrical parameters for a variety of data rates.
(2) *FC-1 or the transmission layer:* FC-1 defines the transmission protocol, including serial encoding and decoding rules, special characters and error control. The information transmitted over a fibre is encoded 8 bits at a

time into a 10-bit transmission character. The primary rationale for use of a transmission code is to improve the transmission characteristic of information across a fibre.

(3) *FC-2 or the signaling/framing layer:* The signaling protocol (FC-2) level serves as the transport mechanism of Fibre Channel. The framing rules for the data to be transferred between ports, the different mechanisms for controlling the different service classes and the means of managing the sequence of a data transfer are defined by FC-2. The different building blocks defined are ordered sets or 4-byte transmission words containing data and special characters, frames or information to be transmitted. The sequence is formed by a set of one or more related frames, to be transmitted uni-directionally. The exchange is composed of one or more nonconcurrent sequences for a single operation. The protocol is related to the service offered by Fibre Channel.

(4) *FC-3 or the common services layer:* The FC-3 level of the FC standard is intended to provide the common services required for advanced features such as striping (using multiple ports to send a single piece of information), hunt groups (the ability of more than one port to respond to the same alias address) and multicasting.

(5) *FC-4 or the applications interfaces layer:* FC-4, the highest level in the FC structure, defines the application interfaces that can execute over Fibre Channel. It specifies the mapping rules of upper-layer protocols using the FC levels below it. Fibre Channel is equally adept at transporting both network and channel information and allows both protocol types to be concurrently transported over the same physical interface.

5.4.1.2 Threats to SANs

In this subsection, we will describe the possible attacks on a SAN. We assume that Fibre Channel is used as the communication infrastructure for the SAN. It is to be noted that most of the attacks discussed here can be recreated in some form or other for topologies which do not use Fibre Channel [24].

Session hijacking

In session hijacking attacks, a malicious attacker takes control of a valid session between two trusted entities. Several weaknesses in the Fibre Channel protocol allow the attacker to launch such attacks. The first is due to clear text messaging: this is required due to performance constraints; any type of encryption significantly reduces the overall throughput of the system. The second is due to the lack of authentication in the basic Fibre Channel protocol. The third is due to the predictable frame sequence: an attacker can easily hijack a session by sending frames to an authorized node with the correct sequence ID and sequence count; this is possible since the sequence ID never changes in a session and the sequence count increments in a predictable manner. However, due to the heavy traffic volume and data speed these attacks are not easily attainable and may have limited attack potential.

Address spoofing

There are three types of login which are important in a Fibre Channel. They are Fabric Login (FLOGI), Port Login (PLOGI) and Node Login (NLOGI). The first allows a node to log in to the fabric and receive an assigned address from a switch. The node sends its information (node name, port, service parameters) to a well-known fabric address; it uses a static 24-bit address to do this as it has not received its own address. The FLOGI will be sent to the well-known fabric address and the FC switches and fabric will receive the FLOGI at the same address. After a switch receives the FLOGI, it will give the port associated with the node a 24-bit address that pertains to the fabric itself. After the node has completed the FLOGI and has a valid 24-bit fabric address, it will perform a PLOGI to another well-known address to register its new 24-bit address with the switch's name server. The switch then registers that 24-bit fabric address, along with all the other information submitted, with the name server and replicates that information to other name servers on the switch fabric. A NLOGI is somewhat similar to a FLOGI, but instead of logging in to the fabric, the node would log in to another node directly (node-to-node communication). A malicious node can spoof the 24-bit address of a valid zone and send the PLOGI request to the well-known address. This will allow the malicious attacker to transfer any trusted information to another location, to be used for malicious purposes. Using the same techniques, a man-in-the-middle attack can be performed. In such an attack, the attacker obtains spoofs using the above method and gets into the conversation between trusted nodes A and B. To node A, the malicious attacker acts as node B, while to B, the attacker acts as node A. Address spoofing can be really dangerous and can be exploited effectively by attackers.

Zone hopping

Switches are currently the only entities to grant/deny access to the SANs. Access is granted mainly based on world-wide names (WWN), which are the MAC addresses of host-bus-adapter (HBA) or network-interface cards. Switch zoning is a technique which allows specific nodes to access other nodes in the fabric using different zoning policies. To launch a zone-hopping attack, the malicious attacker spoofs a WWN address and subverts the zoning table, allowing an unauthorized WWN to access the information of the spoofed WWN. Without spoofing a WWN, if an unauthorized WWN knows the route to another WWN in a different zone which can be enumerated via the fabric, access will be granted.

LUN masking attacks

Logical unit number (LUN) masking is the process of hiding or revealing parts of the storage disk or LUN to the client node. LUN masking creates subsets of storage within the SAN virtual pool, allowing only designated servers to access them. LUN masking can take place at a client node, the FC switch, the storage node or a third-party storage device. If a LUN masking occurs on the client node using HBA drivers, the malicious attacker can change or view the mask settings if the client does not have specific authentication.

5.4.2   Security in Distributed File Systems

Distributed file systems have evolved over the years from easy-to-use remote file systems to highly-distributed file systems with several independent servers spanning multiple networks.

5.4.2.1   Distributed File Systems Overview

Distributed file systems can be broadly grouped into three main categories: remote file systems, network-attached devices and highly-distributed systems.

Remote file system

In the beginning, remote file systems were simply designed to be as easy to use as local file systems. With few exceptions, the majority of these file systems were more concerned with making data available than with securing it. NFS, the de facto UNIX network file system, and CIFS [25], the de facto personal computer network file system, are the distributed file systems in common use today. They are not appropriate for use on untrusted networks due to their weak security features. Now that the Internet is in general use and vast quantities of data are shared over it, the problem of malicious agents has become more important. The susceptibility of the Internet to malicious agents has been known for a long time, but the increased connectivity to it has increased the exposure to such agents.

Network-attached devices

To increase file-system performance, network attached secure disks (NASDs) [26] are used to allow clients direct access to file data stored in object storage devices. NASDs provide clients with secrets which are used to access the object storage devices and prove to them the ability to request an operation on an object. The communication between client and NASD is both encrypted and authenticated. Although each NASD device does not know how the object it holds fits into the file system, the file server still has full access to the file system metadata and the ability to access all of the file system data.

Highly-distributed systems

OceanStore [27] is another highly-distributed file system. It uses a large number of untrusted storage devices to store redundant copies of encrypted files and directories in persistent objects. Objects are identified by globally unique identifiers (GUID), which are generated in a similar fashion to the unique identifiers in Self-certifying file system (SFS). Each identifier is a hash of the owner's public key and a name. Objects can point to other objects to enable directories. All objects are encrypted by the clients. By replicating the objects among servers, clients can avoid malicious servers deleting their data. The extensive use of replication and public keys make revocation of access and deletion of data difficult to achieve, but it does provide a nice model for a completely decentralized distributed file system.

5.4.2.2 Threats to Distributed File Systems

The threats that are common in distributed systems can be divided into passive attacks and active attacks.

Passive attacks

Passive attacks include analyzing traffic, monitoring unprotected communications, decrypting weakly-encrypted traffic and capturing authentication information (such as passwords). Passive intercept of network operations can give attackers indications and warnings of impending actions. Passive attacks can result in the disclosure of information or data files to the attacker without the consent or knowledge of the user. Examples include the disclosure of personal information such as credit card numbers and medical files.

Active attacks

Active attacks include attempts to circumvent or break protection features, introduce malicious code, or steal or modify information. These include attacks mounted against a network backbone, exploitation of information in transit, electronic penetrations into an enclave and attacks on an authorized remote user when they attempt to connect to an enclave. Active attacks can result in the disclosure or dissemination of data files, denial of service or modification of data.

5.5 Overview of Infrastructure Threats and Vulnerabilities

In this section we will briefly provide an overview of the infrastructure threats and vulnerabilities. We have analyzed the threats based on two main categories: impact and time frame. The importance of a threat can be high, medium or low based on the destruction it can generate. Similarly, threats can also be categorized into: immediate, medium-term and long-term threats. Threats are immediate if they can happen any moment, or rather such attacks are immediately feasible.

Table 5.1 Infrastructure threats.

Attack	Attack category	Impact	Time frame	Remarks
DoS	Network threats	High	Immediate	DoS attacks are the most popular attacks on any networking infrastructure and immediate solutions are needed
DNS		Medium	Immediate	These attacks are important, but not as rampant as DoS
Routing		Medium	Long-term	These attacks are theoretically feasible and dangerous, but can only happen in the long term
Wireless		High	Immediate	This is a high-impact area, as wireless applications and users are increasing
Architecture	Grid threats	High	Immediate	This is an immediate potential threat
Infrastructure		Medium	Medium-term	With the evolution of scavenging grid, this threat will increase
Management		Medium	Medium-term	This threat will also increase with increasing use of grid
Trust		Low	Long-term	Not immediate, will be important when there are more autonomic grid networks
SAN threats	Storage threats	High	Immediate	Threats need to be addressed immediately
DFS threats		High	Immediate/ medium-term	These also need immediate attention. Highly-distributed file systems are still evolving

Medium-term threats are those which can take place in three-to-five years' time. Finally, long-term threats will not take place within five years as either the technology is not mature enough to be attacked or it is not popular enough to generate the amount of destruction that the attackers want.

If we look at the *network infrastructure*, the denial-of-service (DoS) attack is an immediate threat and its impact is very high. It can be easily generated and its potential is huge, as whole enterprises can be shut down, resulting in millions of US dollars of loss of revenue. Similarly, attacks targeting the wireless infrastructure have huge potential, as more and more and more applications are becoming wireless. In addition, the wireless user base is increasing exponentially. Therefore, targeting the wireless infrastructure not only has immediate impact, but the attacks can be enormously destructive.

In *grid computing*, the architectural threats are immediate as well as destructive, as the infrastructure can be subverted maliciously for these purposes. Since most of the current grid infrastructure resides within an organization, the impact of infrastructure attacks are limited and can only be felt in the medium term. Similar argument can be put forward for management issues as well. These issues will gain in importance when the scavenging grid and inter-organization grid grow in prominence.

In the case of storage infrastructure, the threats to SANs are immediate and have destructive potential, as most of the enterprise data are stored there. Similarly, the threats to distributed file systems are also important. However, some of the highly-distributed file systems like OceanStore are still evolving and threats to these will only be important in the medium term

References

[1] Chakrabarti, A. and Manimaran, G. (2002) Internet infrastructure security: a taxonomy. *IEEE Networks*, **16** (6), 13–21.
[2] Houle, K.J. and Weaver, G.M. (2001) *Trends in Denial of Service Attack Technology*, CERT Advisory, Vol. 1.0.
[3] CSI/FBI (2001) Computer Crime and Security Survey, available at http://www.crime-research.org/news/11.06.2004/423, accessed on June 13th, 2008.
[4] Moore, D., Voelker, G.M. and Savage, S. Inferring Internet Denial-of-Service Activity. Proceedings of the 2001 USENIX Security Symposium, Washington, DC, 2001.
[5] Criscuolo, P.J. (2000) *Distributed Denial of Service Trin00, Tribe Flood Network, Tribe Flood Network 2000, and Stacheldraht CIAC-2319*, Department of Energy Computer Incident Advisory Capability (CIAC), Lawrence Livermore National Laboratory, UCRL-ID-136939, Rev. 1.
[6] Web Report (2000) Yahoo on Trail of Site Hackers. Wired.com, available on http://www.wired.com/news/business/0,1367,34221,00.html, accessed on 13th June, 2008.
[7] Web Report (2002) *Powerful Attack Cripples Internet*, Associated Press for Fox News, available at http://www.linuxsecurity.com/content/view/112716/2/, accessed on June 13th 2008.
[8] Computer Emergency Response Team (1998) *Multiple Vulnerabilities in BIND*, CERT Advisory, Nov. 1998.
[9] Eastlake, D. (1999) Domain Name System Security Extensions, RFC 2535, Mar. 1999.
[10] Malkin, G. (1998) RIP Version 2, IETF RFC 2453.
[11] Moy, J. (1995) OSPF Version 2, IETF RFC 1583.
[12] Rekhter, Y. and Li, T. (1995) A Border Gateway Protocol 4, IETF RFC 1771.
[13] Welch, J. and Lanthrop, S. (2003) Wireless security threat taxonomy, *Information Assurance Workshop, 2003*, IEEE Systems, Man and Cybernetics Society, June 2003, pp. 76–83.
[14] Potter, B. (2003) Wireless security's future. *Security and Privacy Magazine, IEEE*, **1**(4), 68–72.
[15] Earle, A.E. (2005) *Wireless Security Handbook*, Auerbach Publications.

[16] Foster, I., Kesselman, C. and Tuecke, S. (2001) The anatomy of the grid: enabling scalable virtual organizations. *International Journal of Supercomputer Applications*, **15**(3), 200–222

[17] Foster, I. and Kasselman, C. (2004) *The Grid 2: Blueprint for a New Computing Infrastructure*, Morgan Kaufman.

[18] Chakrabarti, A. (2007) *Grid Computing Security*, Springer.

[19] Gustafson, D., Just, M. and Nystrom, M. (2004) Securely Available Credentials (SACRED) – Credential Server Framework, IETF RFC 3760.

[20] Mui, L., Mohtashemi, M. and Halberstadt, A. (2002) A Computational Model of Trust and Reputation. The 35th Hawaii International Conference on System Science (HICSS), Maui.

[21] Xiong, L. and Liu, L. (2002) Building Trust in Decentralized Peer-to-Peer Electronic Communities. The 5th International Conference on Electronic Commerce Research (ICECR), Montreal.

[22] X3T9.3 Task Group of ANSI (1993) Fibre Channel Physical and Signaling Interface (FC-PH), Rev. 4.2, October 8, 1993.

[23] Fibre Channel Association (1994) Fibre Channel: Connection to the Future, ISBN 1-878707-19-1.

[24] Dwivedi, H. (2005) *Securing Storage: A Practical Guide to SAN and NAS Security*, Addison-Wesley.

[25] The Open Group (1992) Protocols for X/Open PC Internetworking: SMB, Version 2, September 1992.

[26] Gibson, G.A., Nagle, D.F., Amiri, K. *et al.* (1997) File Server Scaling with Network-Attached Secure Disks. Proceedings of the ACM International Conference on Measurement and Modeling of Computer Systems (Sigmetrics), June 1997, pp. 272–84.

[27] Kubiatowicz, J., Bindel, D., Chen, Y. *et al.* (1999) *Oceanstore: An Architecture for Global-Scale Persistent Storage*, ASPLOS, December 1999.

6

Application-Level Threats and Vulnerabilities

6.1 Introduction

Over the last few years there has been a tangible shift in the targets of attacks from networks/hosts to the applications themselves. Today attackers are increasingly concentrating on exploiting the design and coding weaknesses inherent to applications, which is facilitated by a number of factors, such as: a lack of security focus and awareness in software developers, who end up producing buggy software; the wide availability of public information about security vulnerabilities and exploits; availability of sophisticated free and commercial tools which help in exploiting the weaknesses without requiring in-depth security knowledge; and so on. The National Institute of Standards and Technology (NIST) [1] estimates economic impact due to buggy software in the order of tens of billions of US dollars every year. Gartner [2] estimates that 75% of malicious attacks come from application-layer vulnerabilities rather than network or host vulnerabilities.

There are several popular public sources, like the Open Web Application Security Project (OWASP) [3] and Web Application Security Consortium [4], which publish the latest information pertaining to application security threats and vulnerabilities.

This chapter focuses on some of the more critical application-layer vulnerabilities commonly found in applications. While many of these vulnerabilities are applicable to all sorts of applications, some vulnerabilities are very specific to Web applications. This list is not exhaustive but covers the typical security vulnerabilities that every developer should be aware of. A comprehensive coverage of all possible application security vulnerabilities is outside the scope of this book. The References section provides links to some very detailed treatments of this subject,

Distributed Systems Security   A. Belapurkar, A. Chakrabarti, S. Padmanabhuni, H. Ponnapalli, N. Varadarajan and S. Sundarrajan © 2009 John Wiley & Sons, Ltd

which are must-reads for any developer who would like to develop sensible code that is inherently secure and difficult to hack.

6.2 Application-Layer Vulnerabilities

An application-layer vulnerability is fundamentally either a weakness in the design of an application, or an insecure coding practice which comprimises some of its security requirements. These vulnerabilities are introduced into the code by a variety of factors, such as lack of security awareness among the designers and developers, lack of security check points (reviews) at different stages of the development lifecycle, invalid assumptions about the application deployment environment and so on.

Some of the more commonly found application-layer vulnerabilities are covered in the following sections.

6.2.1 Injection Vulnerabilities

Injection vulnerabilities represent a class of vulnerability which primarily result from improper input validation or putting excessive trust on the input of the user. Specifically, the injection vulnerability is one where the user-supplied input is used as part of a command execution or query. If the user-supplied data is not properly sanitized and validated, a malicious user can inject carefully-crafted malicious data to change the semantics of the query or command, which leads to an undesirable consequence. Injection vulnerabilities can have a severe impact and result in loss of confidentiality/integrity, broken authentication and arbitrary command execution. In the worst case, they can result in complete compromise of the underlying application or system.

There are several attacks which exploit the injection vulnerabilities; these include SQL injection [5–8], LDAP injection [9], XPath injection [10], blind injection and so on. A complete treatment of all these attacks and their variations is beyond the scope of this book; the interested reader can refer to [5–10] for more details. However, we do cover some injection attacks below as examples of the fundamental approach followed in exploiting injection-based flaws.

6.2.1.1 SQL Injection

SQL injection occurs when an application does not validate and sanitize user input sufficiently well to discard bad data and uses it directly to construct SQL commands. In such a case, an attacker can inject one or more malicious SQL command as part of the input, to change the meaning of the original SQL command from that intended.

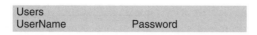

Figure 6.1   Sample table to illustrate SQL injection attack.

The use of SQL injection allows the attacker to insert, delete and modify data, bypass authorization and run arbitrary SQL commands. The impact can be devastating if the database is accessed through a high-privilege user account.

Let us visualize SQL injection with an example. Assume the user credentials are in a table called 'users', as shown in Figure 6.1, and the following SQL statement is written to access user credentials:

```
SELECT * FROM Users WHERE UserName = '" + strUserName + " ' AND
    Password = '" + strPassword + "'"
```

Here assume 'strUserName' and 'strPassword' are the UserName and Password values entered by the user in the login screen. If the user enters the correct UserName and Password, this statement returns the corresponding row. Assume the authentication logic checks to see if a record is retrieved or not, and if a record is returned by the query, the user is authenticated.

However, assume an attacker enters the UserName as " OR 1 = 1 --" in the UserName field and leaves the Password field blank. With this input the above SQL statement is built as:

```
SELECT * FROM Users WHERE UserName = '' OR 1=1 -- ' AND
    Password = ''
```

As you can see above, this input effectively changes the meaning of the original SQL statement. The '--' is treated as comment by most database servers; therefore, the following statement effectively gets executed on the database:

```
SELECT * FROM Users WHERE UserName = '' OR 1=1
```

In the above SQL statement, the OR condition 1 = 1 always evaluates to true and as a result this SQL statement returns all rows in the users table. This allows the attacker to login to the application as the first name in the users table without requiring any password.

The same logic can be extended to gain unauthorized access to, and modify and delete, tables/records in the database. A complete treatment on how the attackers detect the presence of the SQL injection vulnerabilities is out of the scope of this book and interested readers can refer to [5–8] for more details on SQL injection attack.

Since SQL injection is primarily caused by lack of input validation, prevention
is possible by using parameterized stored procedures over and above sanitizing
and validating the input for the 'good' input and denying all other input. More
details on the specific solutions are provided in Chapter 10.

LDAP, XPath, XSLT and blind injection, as well as XSLT, are variations of
the SQL injection where the attacker tries to inject the appropriate commands to
change the logic.

6.2.1.2   LDAP Injection

The Lightweight Directory Access Protocol (LDAP) is the protocol most commonly used to access information from directories which primarily organize and
store information in hierarchical (tree) structure. In today's enterprises, directories
are the common storage for user data and other most frequently-accessed data.
For example, directories are used in organizations to store user credentials, user
attributes and information about other resources like printers, fax machines and so
on. The users are typically authenticated to LDAP directories. These directories
often contain sensitive data and are a hot target for attackers.

LDAP injection vulnerability is where the user-supplied input is directly used
without proper validation to construct an LDAP query statement. An attacker can
exploit this vulnerability by supplying carefully-crafted LDAP statements as part
of the input, changing the semantics of the LDAP statement. If successful, the
LDAP injection vulnerabilities can cause information disclosure, bypassing the
authentication and access-control mechanisms and so on. Interested readers can
refer to [9] for a more comprehensive treatment of these attacks.

6.2.1.3   XPath Injection

XPath is the language used to refer to parts of an XML document. XPath is
either used directly to retrieve specific portions of an XML or in conjunction with
XQuery or XSLT transformations.

XPath injection is where the user-supplied input is directly used without proper
validation to construct XPath queries. An attacker can inject XPath constructs as
part of the user input, changing the semantics of the originally-intended XPath
statement. If executed, this results in advantage being given to the attacker and
compromises the security of the system. Interested readers can refer to [10] for a
more detailed explanation of XPath attacks.

6.2.1.4   Blind Injection

There are many variations on how an attacker can exploit the SQL, LDAP
and XPath injection vulnerabilities. One specific variation of interest is Blind
injection, where the attacker is not aware of the underlying structure of the

SQL/LDAP/XPath statement but can determine the complete structure of the underlying statement and ultimately exploit the system. Interested readers can refer to [6] for more details on Blind injection attacks.

6.2.2   Cross-Site Scripting (XSS)

Cross-site scripting (XSS) attacks are made possible by injection vulnerabilities in the application code. Although XSS is a specific attack belonging to the general class of injection vulnerabilities, it is covered here separately due to its severe impact on applications. XSS typically occurs in sites which allow user input, and echoes this back to the client without proper validation. In XSS attack, an attacker injects malicious client-side script as part of the input data, which is subsequently echoed back to the client browser. As the malicious script is served through the genuine server, the browser trusts it and executes it with the security privileges applicable to the site, which means the script can access the cookies set by that site and steal or manipulate the data silently to a remote server, read/alter the page content and so on.

XSS is one of the more prevalent exploits on Web applications and is common in applications like search engines, where the user-entered search key words are echoed; message forums, where the user-submitted messages are seen by everyone; error messages, which echo the user-entered data; and so on.

Consider an on-line message forum portal that is vulnerable to XSS attack. Users of the messaging portal typically post messages or view the messages posted by other users. Assume an attacker with malicious intent enters a message like '*This is a bad post script alert('XSS')*'.

Any other user of the portal who views that message receives '*This is a bad post script alert('XSS')*' and the user's browser interprets the *script* tags as JavaScript and executes it to display an alert as 'XSS'. Once such a vulnerability is discovered in a page, the attacker can write more malicious scripts to steal cookies, session tokens and so on. Though we have demonstrated here an example of this attack that uses JavaScript, an attacker can inject any client-side executing script like HTML, Flash and so on. XSS attacks can be launched to disclose confidential data, or to break the access control restrictions of the application.

There are primarily two types of XSS attack that are possible: persistent and nonpersistent (reflective). In a nonpersistent XSS attack the attack is carried out as the victim is interacting with the application. The attacker typically sends a specially-crafted link to the victim and persuades him to click on it. An unsuspecting victim, upon clicking the link, is presented the page with the said XSS vulnerability by the unsuspecting genuine server. This leads to the vulnerable script being executed on the client's browser. In a persistent XSS attack, the attacker typically posts the maliciously-crafted code to the server, which stores it in a repository (e.g. files, database) and subsequently presents it to other users when they request that content. A typical scenario is a message forum that is vulnerable to this kind of attack. An attacker posts a malicious message to the forum, and subsequently all other users are affected when they view that message. For a detailed treatment of the XSS attack and exploit codes, see the References section.

XSS attacks are primarily caused by echoing the user input as part of the response as it is entered, without proper validation. Interested readers may refer to [4, 11] for more details on XSS attacks. Remediation strategies for XSS attacks include proper validation of input, encoding the input data when it is echoed back and so on. More details on the specific solutions are provided in Chapter 10.

6.2.3 Improper Session Management

Session management flaws are very common in Web applications. HTTP is a stateless protocol, which means that a Web server cannot relate a series of requests as coming from the same user. Since this is a necessary ability for stateful application, the concept of user sessions was introduced to Web applications. There are predominantly three types of technique used to support sessions: hidden form fields, URL rewriting and cookies. Typically in all these techniques, a session identifier that identifies a user session on-server will be exchanged in every subsequent request and response between client and server. The design and secure implementation of this session management logic is often the responsibility of the application and is critical for the security of the application.

Any flaws in the session management architecture and its implementation result in broken authentication and access control and provide inadvertent advantage to the attacker. For this reason, session identifiers and session-management techniques are often the targets for attackers. In spite of the critical role of session management in Web applications, it is astonishing to note that many Web applications continue to be vulnerable to improper session management flaws, resulting in compromised systems.

Different attacks are possible, depending on how the session management is designed and implemented. The session IDs can be subjected to interception, prediction and brute-force analysis attacks, and any flaws in the session management architecture can lead to session hijacking, session fixation, replay and man-in-the-middle attacks.

As obtaining a valid session ID often allows an attacker to gain direct access to the application without any credentials, session IDs are the prime target for attackers. Session ID interception attack exploits the weakness in session management design, where the session identifiers are exchanged over an insecure communication channel. This allows the attacker to use a wide variety of tools to intercept the traffic and gain access to the session ID being exchanged. The spoofed session IDs can be used to hijack the victim's session. Use of encrypted sessions or secure sockets-layer (SSL)/transport-layer security (TLS) is a recommended way of discouraging session ID interception attacks. In session-prediction attack, the attacker successfully predicts the valid session identifiers. This can be due to a weakness in the design and implementation of how the application generates the session identifiers. For example, the use of sequential session IDs in proprietary session ID generation modules makes the prediction of session identifiers an easy task for attackers. Cryptographically-secure random number generators, along with random seeds, are a more secure choice, since they make the prediction of session identifiers a tough job for attackers. Brute-force analysis of session IDs is where an attacker tries all possible combinations of session IDs in an exhaustive manner. Choosing sufficiently large session identifiers relative to the number of simultaneous sessions present at any given point of time helps in countering the brute-force analysis attacks on session identifiers.

In a session fixation attack [12], the attacker utilizes a design or implementation flaw in session management logic of the application to fix the session ID to a value of their choice and subsequently lures the victim to use that session ID for login. Session fixation is possible when the session identifiers are either chosen by the user or set by the server. For example, consider the case of an online banking application that has a session management flaw.

Assume that the bank's online portal creates a session identifier for the user when they first connect to the banking site using their browser, even before they have logged in using their credentials. Further assume the bank uses the same session identifier to continue to maintain the user's session after the event of successful user login. This is an ideal scenario for an attacker to launch a session fixation attack. The attacker connects to the banking site first. The banking site creates a session and passes the session identifier to the attacker's browser (say, as a URL parameter). The attacker intercepts this session identifier and creates a carefully-crafted link that points to the bank's login page, where their own session ID is already set as a URL parameter, and lures the legitimate user to click on it. If the legitimate user clicks on the link and connects to the banking site, the banking site simply reuses the session identifier and supplies the login screen in the context of that session. Now if the legitimate user logs in to the bank's site using their credentials, they share a session ID with the attacker. The attacker can then bypass the login screen and access the bank's site as the legitimate user.

There are several other techniques with which an attacker can get access to a valid session identifier belonging to a legitimate user. For example, consider a banking site that has XSS vulnerability. Further assume the bank sets the session identifier as part of the cookies on the client's browser. An attacker who is aware of the XSS vulnerability can create a message with a carefully-crafted link and lure the user to click on it. When the user clicks on the link, the script injected by the attacker can silently steal the cookie information and send it to them. Now they have access to the legitimate session identifier and can connect to the application without having to prove their credentials.

In conclusion, there are many variations in how the design and implementation flaws in session management can be exploited by an attacker to gain access to a system. Session management forms a critical aspect of Web application design and has to be carried out with security in mind. Interested readers may refer to [12] for more details on session fixation attacks. Several solutions and best practices for secure session management are discussed in Chapter 10.

6.2.4 Improper Error Handling

Error handling is one of the rather frequently ignored aspects of secure application design. Improper error handling vulnerabilities in applications lead to disclosure of sensitive information like database details, platform version details, SQL statements, stack traces and so on to attackers. Many times the error messages generated by applications contain lots of useful information that inadvertently helps an attacker instead of being helpful to the user. Sometimes error messages differ for the same error conditions but different versions of the product or for different input. All this leads to leakage of important information which helps the attacker know more about the application environment and thereby launch specific attacks against the known vulnerabilities of the application.

Attackers inject malformed inputs to specifically force applications to fail. This helps the attacker either in getting the sensitive useful information that is leaked through error messages or, in some cases, in causing denial of service (DoS).

For example, consider an application that throws different error messages during the login process depending on the particular sequence of user actions. Assume that if an invalid login name and password is supplied to the login page, after validation it returns an error saying 'invalid login user name or password'. This by itself is a 'good' error message because it does not reveal exactly which of the ID or password is actually incorrect. Suppose the attacker has access to one valid login user name and he enters it (along with an incorrect password) in the login screen. If the application now returns a different error message (one that says 'invalid password', for example), it provides a sufficient hint to the attacker that he now possesses a valid login user ID for sure and only has to guess the right password to get access to the application.

Many times, the error conditions are not properly caught in the application layers, and this results in displaying the stack trace (corresponding to the instant when the exception was actually thrown) directly to the user. The stack trace is typically provided primarily for the developers, to help them debug the application; hence, it often contains sensitive details like SQL commands, line numbers, the technology stack being used and so on. This information is of no use to the end user, but proves to be of immense help to attackers. For example, because of some SQL query error a stack trace might return the database name, table name, column name and the specific reason why the query failed. If this is displayed to the attacker, they can carefully craft messages to determine the entire database schema and subsequently launch attacks. A more detailed example of how an attacker can perform SQL injection attacks with the aid of stack traces is given in the References.

To prevent information disclosure or leakage, security issues are to be thoroughly considered while designing the error handling strategy of an application. Enough care must be taken to ensure that error messages are consistent and only contain the minimum information required to reveal the possible cause. Also, care must be taken to see that error messages are consistent across different versions of the same software. A centralized exception handling is a recommended approach. More information about the best practices for secure exception handling is given in Chapter 10. Interested readers may refer to [13, 14] for detailed discussions of the issues related to improper error handling.

6.2.5 Improper Use of Cryptography

The use of cryptography in applications often leads to a false sense of security among the developers. However, the reality is that cryptography is not the solution for every security problem; it only provides solutions for specific problems like authentication, data secrecy, nonrepudiation and so on. For example, coding bugs that lead to security breaches cannot be addressed by cryptography. The other huge misconception is that all crypto implementations are equally secure. It is very dangerous to assume that all cryptography is equal. Different cryptographic algorithms have different security strengths. It can cost one dearly if cryptography is not implemented safely. Defective implementations of crypto algorithms, wrong assumptions made about randomness sources, bad use of crypto algorithms in applications, in-house developed algorithms and so on all prove costly and often end up lulling an organization into a false sense of security.

For example, developers often believe that a crypto algorithm that has been developed in-house and hence is not known to the public at large is safe from attackers. That this is really not so has been proved many time. Given enough resources (time, computing power), any crypto algorithm can be broken. For example, RC4 was initially a trade secret, but in September 1994 a description of it was anonymously posted to the Cypherpunks mailing list. It was soon posted on the sci.crypt newsgroup, and from there it made its way to many sites on the Internet. Security through obscurity might help in certain cases to raise the bar for an attacker, but it should not be the only defense to prevent an attack. The strength of an algorithm lies in the secrecy of its key, not in the secrecy of algorithm detail. Proprietary algorithms are often not well tested and analyzed. Therefore, it is recommended that published algorithms, which are time-tested against stringent security attacks and have been subject to cryptanalysis by several cryptographers and attackers, always be used unless there is a good reason for not doing so.

Cryptographic operations like cryptographic key generation, password generation and so on require the use of random numbers. The characteristics for random numbers to be used in crypto functions are stringent; otherwise the functions would be subject to multiple weaknesses which could be exploited by attackers. The important characteristics for crypto random numbers are that there should be even distribution, the random numbers should be highly unpredictable, and the random number range should be sufficiently high. Many of the common random algorithm implementations do not fulfill these requirements and their outcomes are reasonably predictable. If developers use these insecure algorithms to generate keys, they are highly predictable and it is easy for attackers to break the system. It is critical to look at the documentation of the random generator libraries to see if they support crypto random number generation algorithms. Similarly, it is always better to check the documentation on the source of randomness for the algorithms.

Symmetric cryptography is popularly used to encrypt any sensitive data while it is in transmission or persistent storage. This involves using the same key for both encryption and decryption, and sharing the key between the sending and receiving parties. Frequently, developers of an application hardcode the keys in source code, assuming it is not possible to read them from binary executables. However, there are plenty of methods and tools by which the hardcoded keys can be readily extracted from executables. Besides this information leakage, hardcoded keys make the task of changing the keys at regular intervals very difficult. Similarly, storing sensitive keys in insecure storage may cause leakage of those keys.

Oftentimes passwords are used to generate keys and if the passwords are not chosen carefully, the effective password bit length may be much less than that required, and this can cause the production of predictable keys.

Also, using the right crypto algorithms for the right purpose with the right key lengths is crucial for the secrecy of the data. For example, use of broken algorithms like MD5, DES and so on for cryptographic computations must be avoided. Similarly, using 512-bit RSA with 512 may be considered weak; at a minimum, 1024-bit RSA should be used.

More details on the common pitfalls and best practices in the use of cryptography are given in [13–15].

6.2.6 Insecure Configuration Issues

Applications are only as secure as their weakest link. Even if security best practices have been strictly followed while implementing an application, any single mistake by administrators in configuring and deploying it may completely jeopardize the entire security of the system.

Often application developers bundle their application with a lot of features set to their default configurations, which may have security implications if they are not fine-tuned in the production environment. Releasing applications with default passwords for administrators is a good example of such vulnerability. These default passwords are generally well known, and the first thing attackers try to do is to use these default passwords to access administrative accounts.

Similarly, requiring the application to run with more privileges than required may provide an opportunity for attackers to abuse these privileges.

Most of the time the configuration files will store the sensitive data in plain text. Database connection strings and database user credentials are typical examples of this. This makes the configuration files a tempting target for attackers. If they are not adequately access-restricted, they can be exploited by any insider with malicious intent who has access to them. Also, often the user credentials are transmitted as plain text in the internal network. This makes the application vulnerable to insider attacks, as anyone who has access to the network can use network sniffers to learn the plain passwords. It is good practice to use encryption to protect the sensitive data in configuration files. Similarly, it is good practice to use a secure channel to transmit the sensitive data, even with an organization's internal networks.

Many applications come with administration interfaces that help administrators configure them and their security features. If care is not taken to access-restrict these interfaces for only privileged administrative accounts, they can be abused by insiders. Similarly, if these interfaces can be administered remotely, a lot more care has to be taken to use a strong form of authentication for remote administrators. Interested readers may please to [13, 14] for more discussion on configuration-related security issues.

6.2.7 Denial of Service

Availability to genuine users when they need access is a key requirement for any application to be successful. A DoS attack aims at making an application or service unavailable to users. There are multiple ways in which an application and its services can be attacked, which may range from physical attacks on hardware infrastructure, through attacks exploiting the weaknesses in the underlying network communication protocols and attacks exploiting the vulnerabilities in application code, to attacks exploiting weaknesses in the business processes supporting the application. An application-layer DoS attack is one that exploits the software design and implementation vulnerabilities to make a service unavailable to its users. DoS attacks severely impact the confidence of users in a company, resulting in brand loss and direct-revenue loss.

Application layer DoS attacks are normally very difficult to detect and prevent as often the bandwidth consumed by the application layer DoS attacks is small and indistinguishable from normal traffic. These attacks typically take advantage of the implementation and design flaws in the application and its business process layer. Attackers analyze the application for any possible bottlenecks and try to exploit them.

Attackers can analyze the application to see possible vulnerabilities, like buffer overflow, SQL injection and so on, and can exploit them to cause DoS. Similarly, designers have to take extra care when allowing access to certain functionality by users. For example, assume a report-generation functionality which consumes a lot of computing cycles is exposed to users. If users are not forced to authenticate to the system before accessing this report functionality, it can be abused by attackers to cause DoS, as the attackers can cause the system to generate lots of reports which are, obviously, not intended for consumption. Allowing only authenticated users to access the resource-intensive functionality is often a best practice to lower DoS attacks.

Similarly, attackers can abuse application- and process-layer vulnerabilities. For example, assume login functionality without an account lockout policy, which allows users any number of authentication attempts. This can be abused by attackers to launch dictionary-based attacks or brute-force attacks, which try all possible combinations of usernames and passwords. But if an account lockout policy is implemented to prevent a certain number of invalid authentication attempts, this can further be abused by attackers to lock-out the accounts of targeted individual users. Attackers can obtain valid usernames through different means, including social engineering attacks, guessing of usernames/passwords based on information available about particular users and so on.

Fully-automated features, if they are not implemented with security in mind, can sometimes cause DoS. For example, assume a Web-based e-mail portal that allows users to register their own profiles and create an account without any manual intervention. Such a system can very easily be exploited, by automating attacks through a script to make a large number of false registrations.

For a more comprehensive treatment of application-layer DoS vulnerabilities, please refer to [16]. As DoS attacks have a severe impact, application designers and developers should carefully consider their possibility and understand how they can be mitigated. More information about the best practices for addressing application-layer DoS attacks is given in Chapter 10.

6.2.8 Canonical Representation Flaws

Canonical representation of data means representing the data in its simplest, most direct and basic form. Often multiple formats exist in computer systems to represent the same data. For example, the same data can be represented using different character-encoding schemes like ASCII, Unicode (UTF-8, UTF-16 and UTF-32), ISO 8859-1 and so on. A back slash '\' character can be represented in different forms in UTF-8, as <5C> or <C1 9C> (in its shortest-form representation). Similarly, in some operating systems the file names and directory names can be represented in multiple ways. For example, MS DOS represents files with its 8.3 notation, where mysecretdata.txt and mysecr~1.txt may represent the same file.

In a well-designed application, all input will be checked for potential security implications by some sort of input-validation routine. After the input has been scrutinized, the application components can trust it to be secure. However, if the input validation fails to take care of the multiple possible representations of the same data and takes a decision based on the noncanonical representation of the data, it may lead to severe security flaws, such as spoofing, information disclosure and broken access control threats.

A comprehensive treatment of canonical representation issues is out of the scope of this book; interested readers may refer to [13–15, 17].

6.2.9 Overflow Issues

As a general concept, overflow is said to happen when data is written to a buffer without insufficient-bounds checking and is therefore allowed to overflow into (and hence, corrupt) one or more locations adjacent to the allocated buffer. This has been known to be exploited to manipulate a running program into changing its behavior in a way that benefits the attacker.

The stack overflow attack involves writing more data to a fixed-length buffer on the running program's call stack than is actually allocated for that buffer. What this essentially implies is that the program, which may be running with special privileges, has its stack buffer injected with 'data' of the attacker's choosing. Given that the stack also contains the return address of the currently executing procedure, all the attacker has to do is to inject the executable code on the stack and simultaneously overwrite the 'most recent' return address with the starting address of this piece of code. As soon as the currently executing function returns, it will pop the return address from the stack and begin executing instructions starting from that address (which is essentially the attacker's code).

Broadly, there are two ways in which such exploitation can be prevented. If we are able to detect that an overflow has happened, we can 'stop' the instruction pointer from jumping to the (corrupted) return address that is present on the stack. As an alternative, we can, as a general rule, enforce a memory policy on the stack memory region to disallow execution from the stack altogether.

The former method essentially entails placing a random integer in the memory location just before the return address of the currently executing function. The technicalities of the way stack overflow happens effectively ensure that touching the return address corrupts this integer as well. A simple check on the contents of this specific location, just before the executing routine jumps to the return address, is enough to confirm that an overflow has happened.

The second method, Data Execution Prevention, entails setting the NX ('No eXecute') or the XD ('eXecute Disabled') flag – assuming the CPU supports it – to mark the specific pages in the memory that contain the stack (and the heap, if need be) as readable but not executable. As a result of this marking, any attempt to execute the code that the attacker may have managed to inject into the stack will lead to a kernel exception.

6.3   Conclusion

There is a large amount of information available in the public domain on common application security vulnerabilities. At the same time, there are also plenty of commercial and free open-source tools available that help in exploiting those weaknesses. Often, application security is an afterthought and not built into the application as an integral step of the software-development lifecycle. This results in buggy software from a security perspective. These reasons make the application and process layers tempting targets for attackers. Designers and developers must pay close attention to ensure that security is built into the application from the ground up. This chapter has tried to cover some typical application security vulnerabilities and the corresponding possible attacks. It is very important for all application developers and designers to be aware of these common application security threats, so that they can build appropriate countermeasures into their applications.

References

[1] National Institute of Standards and Technology, http://www.nist.gov/
[2] The Economic Impacts of Inadequate Infrastructure for Software Testing, http://www.nist.gov/director/prog-ofc/report02-3.pdf
[3] Open Web Application Security Project (OWASP), www.owasp.org
[4] Web Application Security Consortium Threat Classification, http://www.webappsec.org/projects/threat/classes/cross-site_scripting.shtml
[5] Spett, K. *SQL Injection: Are your web applications vulnerable?*, Technical report, SPI Dynamics, Inc., 2005.
[6] SPI Labs, *Blind SQL Injection*. SPI Dynamics. 2005, accessible spidynamics.com/whitepapers/Blind_SQLInjection.pdf.
[7] Anley, C. *Advanced SQL Injection in SQL Server Applications*, Next Generation Security Software Limited, 2002.
[8] OWASP *SQL Injection*, http://www.owasp.org/index.php/SQL_injection
[9] Faust, S. *LDAP Injection – Are Your Web Applications Vulnerable?* a white paper from SPI Dynamics, 2003.
[10] Xpath Injection, http://www.webappsec.org/projects/threat/classes/xpath_injection.shtml, http://www.owasp.org/index.php/XPATH_Injection
[11] Spett, K. *Cross Site Scripting – Are Your Web Applications Vulnerable?* a white paper from SPI Dynamics, 2005, accessible http://spidynamics.com/whitepapers/SPIcross-sitescripting.pdf.
[12] Session Fixation Vulnerability, http://www.acros.si/papers/session_fixation.pdf
[13] OWASP, A Guide to Building Secure Web Applications and Web Services, 2.0, Black Hat, 2005.
[14] *Improving Web Application Security - Threats and Countermeasures*. Microsoft Press, September 2003.
[15] Howard, M. and LeBlanc, D. *Writing Secure Code*, 2nd edn, Published December 2002.
[16] de Vries, S. Application Denial of Service (DOS) Attacks. A Corsaire White Paper, April 2004.
[17] Security Considerations for the Implementation of Unicode and Related Technology, http://unicode.org/reports/tr36/tr36-1.html

Further Reading

Howard, M. and LeBlanc, D. *Writing Secure Code*, 2nd edn.

7

Service-Level Threats and Vulnerabilities

7.1 Introduction

The notion of service-oriented architecture (SOA) is predicated upon the concept of well-encapsulated bundles of software accessible over networks via well-defined open implementation independent standards-based interfaces. The openness and implementation independence are a source of additional complexity in terms of security requirements. The adoption of Web services enables business organizations to improve business process efficiency by reducing cost and time, and to gain expansive business opportunities. Enterprises have been able to create service wrappers over their huge existing legacy code base. This has enabled them to bring the old legacy systems to the forefront of dynamic process integration using technologies like BPEL and real-time information dissemination. The application of integrated communication makes it possible for business organizations to gain a wider group of collaborative business partners, customers and services, which eventually brings them into new competitive service markets.

While Web services have really opened up the possibilities of intra- and inter-enterprise integration and collaboration, they have also opened up additional threats and vulnerabilities. To add to the complexity, Web services are built at the edge of the enterprise ecosystem, and this, coupled with the fact that they are composed of human-readable message interactions, leaves them extremely vulnerable to attacks.

This chapter focuses on some of the most commonly identified threats and vulnerabilities that exist for service implementations. While many of these vulnerabilities are omnipresent across the various layers, like applications and infrastructure, they represent themselves in the service layer as mutant manifestations.

Distributed Systems Security   A. Belapurkar, A. Chakrabarti, S. Padmanabhuni, H. Ponnapalli, N. Varadarajan and S. Sundarrajan © 2009 John Wiley & Sons, Ltd

This list is not comprehensive, but addresses most typical security vulnerabilities that should be in the knowledge base of service designers.

7.2   SOA and Role of Standards

SOA is primarily predicated upon standards-based communication between services, hence security needs to be addressed with standards in mind.

7.2.1   Standards Stack for SOA

Web services represent the most popular form of SOA, and are based on the idea of a stack of standards for various facets of distributed computing, right from message layer to higher business process layers, with nonfunctional layers running right through. While there exist several WS-* standards constituting this stack, we show a simplified version in Figure 7.1 to facilitate further elaboration of relevant security issues for Web services and SOA. In particular, in Chapter 11, we shall elaborate on the security standards for SOA.

In Figure 7.1, the core Web services standards are illustrated. At the bottommost layer, XML represents the lingua franca of all communications in SOA. All Web services standards are defined in XML, and mandate usage of XML-based documents in communications. Above the XML layer, the XML schema provides for a universal grammar for data types, independent of the implementation language and platform. The Simple Object Access Protocol (SOAP) leverages XML schema to provide for standards-based packaging of messages sent between service providers and consumers, irrespective of the underlying transport protocol (e.g. HTTP, SMTP, HTTPS, etc.). While SOAP is able to provide standards for messaging, the Web Services Description Language (WSDL) provides a standard for describing the communication and location details of a service, to be advertized to all service consumers to help them get access to all the details necessary to

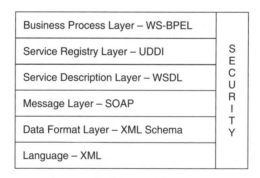

Figure 7.1   Standards stack for SOA.

carry on a service invocation. Universal Description Discovery Integration (UDDI) provides an XML-based standard for service registries, with standards for service publishing and searching services. WS-BPEL is an emerging standard for business process descriptions leveraging orchestrations of Web services. The standards for nonfunctional requirements, including security, which are cross-cutting across all layers, again leverage XML. There are multiple security standards for SOA, addressing the diverse SOA security requirements. We shall study these standards in Chapter 11, as standards are a crucial element of the solutions addressing SOA security issues.

7.3 Service-Level Security Requirements

Traditional security technologies are by and large inadequate to secure Web services. Boundary-based network security technologies (e.g. firewalls, IDS) are inadequate because Web services operate using the underlying HTTP protocol. Firewalls are usually configured with very little restrictions for HTTP. They generally also allow HTTP tunneling for RMI/EJB objects. While certain types of firewall support filtering of HTTP content and some can support filtering of SOAP, filtering all XML messages with traditional firewalls can prove to be expensive and a performance drag, and reduce their throughput. Web service specifications allow for messages to be passed via intermediaries rather than simply by point-to-point communication. SSL/TLS is inadequate for a number of possible SOAP transactions because SSL/TLS is designed to secure transactions between individual Web services – and will not protect against an intermediary performing a man-in-the-middle (MITM) attack. The openness and interoperability features also open up new avenues of attacks.

SOA necessitates a rethink of some of the core security requirements, while creating additional requirements owing to the open network-based application invocations via XML-based standards. Further, today's Web service application topologies include a broad and diverse combination of mobile devices, gateways, proxies, load balancers, demilitarized zones (DMZs), outsourced data centres and globally-distributed dynamically-configured systems. The primary concerns in SOA implementation include the following [1]:

7.3.1 Authentication

In SOA, like in any Web-based application, an invoker must be authenticated to establish genuineness of credentials. Diverse authentication mechanisms (including authentication via username/password, Kerberos, NTLM, Certificates from Trusted CAs, etc.) need to be provided for in Web services. Additionally, multiple applications with heterogeneous authentication systems will need to interoperate with one another. Further, with Web services it might be necessary to accept credentials from outside an enterprise; for example, from a business partner.

Standardization of formats for sharing authentication information hence becomes an absolute necessity.

7.3.2 Authorization and Access Control

In SOA implementations, since requestors invoke operations, beyond the basic access control models for accessing IT resources, there is a need to specify operation (method)-level access privileges (like permission to execute, privileged user to the service, etc.). Also, it might be necessary for authorization and access control information to be updated dynamically, during a multihop service invocation. Further, since SOA is applied in a business context, there is a need to allow for specification and validation against complex policies for access (business policies, security policies, etc.). Standards for interoperation of multiple policy formats become crucial. This also calls for standardization of universal policy languages capable of handling complex formats of requests.

7.3.3 Auditing and Nonrepudiation

In any Web service invocation, the need to be able to know the users, their locations and their access details, is crucial in case of attack by unintended users. This may include capturing of failed invocations of a service, faulty credentials, bad authentications, bad signature occurrences and so on. An administrator should be able to analyze the audit records, in order to predict and nail down attackers.

In addition to being able to capture the fact that a certain piece of information was created or transmitted, auditing in SOA is necessary for nonrepudiation, which requires every service invocation detail to be captured so that at a later time the requestor cannot deny the invocation.

7.3.4 Availability

In any distributed SOA implementation, the service provider should provide a reasonable guarantee that the service will be available to a genuine requestor when needed. This is typically defined and implemented via standard SLA formats which service provider platforms can vouch for and service consumers can agree upon.

A crucial aspect of ensuring availability is detecting and preventing denial-of-service (DoS) attacks, which may be carried out with input of malicious data that appears harmless to conventional solutions like firewalls. This can lead to serious harm to the service provider. The malicious data may be in form of spurious code, sent as part of a Web service request.

In SOA, in addition to malicious data-based attacks, owing to the ease of invocation of services with exposed interfaces, a repeated set of attacks can occur, leading to DoS (and DDoS) and loss of availability. This requires a specialized kind of firewall, with the capacity to diagnose the content of a request and examine methods of invocation in isolation. Code Inspection of service requests can provide a clue about repeated requests and appropriate mechanisms can be used to avoid such attacks

7.3.5 Confidentiality

In SOA, it may be required that a particular portion of an XML document is not understandable by anyone other than the person for whom it is intended, while the remainder of the document is left untouched. This necessitates partial encryption of the document. This is because quite often message invocations in SOA traverse across several intermediaries, with the routing information present in the headers of the messages.

7.3.6 Data Integrity

In SOA, it may be required that a portion of an XML document should not be altered in storage or transit between sender and intended receiver without the alteration being detected, while other portions should be alterable, or in certain cases might even be deliberately altered. For example, in a value-added intermediary extra information might be added to the header of a message, while the body remains untouched.

7.3.7 Privacy

In SOA, well-defined formalisms are needed to use and disclose personal information provided by the service-requesting clients. The legal landscape surrounding data privacy is fluid and is subject to new legislation and interpretation by government agencies and the courts. Organizations creating, managing and using Web services will often need to state their privacy policies and require that incoming requests make claims about senders' adherence to these policies.

7.3.8 Trust

In SOA, applications can come from diverse trust domains, and hence there is a need for flexible message-based trust mechanisms, which can allow for dynamic establishment of trust via message-level information (proof of trust). Hence, trust intermediaries are important in the context of SOA.

7.3.9 Federation and Delegation

Delegation refers to the capability of one service or organization to transfer security rights and policies to another trusted service or organization. Federation is a special case where the transferee service is a peer, not a child. The failure of a service should not let the whole federation collapse; instead a delegation is made to another service.

Single sign-on is a mechanism which allows a user to access different systems and resources without the need to enter multiple usernames and passwords multiple times. Identity is a set of definitely distinguishable attributes or bits of information about a person or business which can be used in authentication and authorization. Federated identity architecture delivers the benefit of single sign-on, but it does not require a user's identity to be stored centrally.

7.4 Service-Level Threats and Vulnerabilities

While Web services have really opened up the possibilities of intra- and inter-enterprise integration and collaboration, they have also opened up additional threats and vulnerabilities. To add to the complexity, Web services are built at the edge of the enterprise ecosystem, and this, coupled with the fact that they are composed of human-readable message interactions, leaves them extremely vulnerable to attacks.

This section focuses on some of the most commonly identified threats and vulnerabilities existing for service implementations. While many of these vulnerabilities are omnipresent across the various layers, like applications and infrastructure, they represent themselves in the service layer as mutant manifestations. This list is not comprehensive, but it addresses most typical security vulnerabilities that are included in the knowledge base of service designers as mandatory. Service-layer vulnerabilities are either weaknesses in the design of a service or deficient coding practices in the service's inherent implementation. They make the service and the underlying application and infrastructure vulnerable in a manner that allows unauthorized users access. These vulnerabilities are introduced into the design and code in the first place by of a variety of factors, such as lack of security awareness among the designers and developers, lack of security check points (reviews) at different stages of the development lifecycle, invalid assumptions made about the service deployment environment and service consumers, and so on.

7.4.1 Anatomy of a Web Service

Figure 7.2 is a representation of all the components and actors involved in a Web service call. Table 7.1 is the mapping of the components to the actors involved in a Web service conversation.

7.4.1.1 Web Services Description Language (WSDL)

WSDL is the XML representation of a service contract. It is an extension of the XML schema to define all the operations that a service provides: the parameters

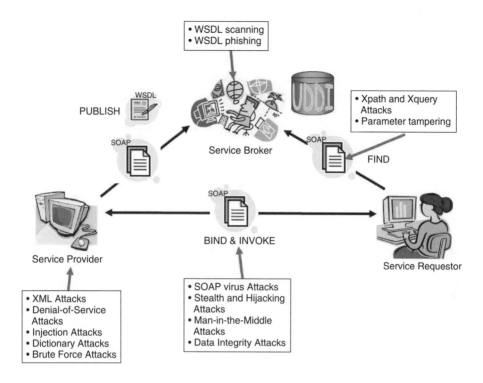

Figure 7.2   High-level services threat profile.

Table 7.1   Mapping of components to the actors involved in a Web service conversation.

Actors	Technology components
Service provider	WSDL, SOAP response message
Service broker	UDDI (WSDL list)
Service consumer	SOAP request message

for the operations. A complete WSDL definition contains all of the information necessary to invoke a Web service.

Hence it also becomes a soft target for attackers, as it offers up a wealth of information. Exposing too much information about a Web service through its WSDL descriptor may provide details about its design and security requirements, and potentially the underlying framework and the host systems.

7.4.1.2   Universal Description Discovery Integration (UDDI)

UDDI is an industry initiative to create a platform-independent, open framework for describing services, discovering businesses and integrating business services.

It is designed as the Yellow Pages, which allows businesses to publish their services. These services can be queried and searched by consumers. UDDI is used to describe your own business services, to discover other businesses that offer desired services, and to integrate with these other businesses.

To develop componentized Web-based services, UDDI provides:

(1) A standardized, transparent mechanism for describing the service.
(2) A simple mechanism for invoking the service.
(3) An accessible central registry of services.

UDDI registries provide details about the purpose of a Web service, as well as detailed information on how to access it. Use of tModels (the technical API) in UDDI allows for description of detailed technical signature including underlying technical structures and interface details describing the behavior. This can be extremely valuable information, which could potentially be misused my attackers. UDDI registries do not provide a robust mechanism for verifying the authenticity of entries. This could allow malicious Web services to be added to the registry and used by other Web services.

7.4.1.3   Simple Object Access Protocol (SOAP)

SOAP is an XML-based stateless, one-way message exchange paradigm. SOAP makes use of an Internet application-layer protocol as a transport protocol. SOAP is silent on the semantics of any application-specific data it conveys, as it is on issues such as the routing of SOAP messages, reliable data transfer, firewall traversal and so on. SOAP however provides the framework by which application-specific information may be conveyed in an extensible manner. Also, SOAP provides a full description of the required actions taken by a SOAP node on receiving a SOAP message. The intention of SOAP was purely to act as a message exchange standard between service provider and service requestor. Hence SOAP specifications do not provide any 'out of the box' mechanisms to perform authentication between intermediaries. There are no techniques to ensure data integrity or confidentiality, either at endpoints or during transit.

7.5   Service-Level Attacks

7.5.1   Known Bug Attacks

There are a lot of known Web service implementations that use popular open-source components and frameworks. The Apache Web server [2] and the Axis Web services engine [3] are probably the most prominent. They also have publicly-available known bugs and issues lists. Even with proprietary products, security consortiums issue warnings against known bugs and security weaknesses. These bugs are generally fixed by the framework or product owners with new releases or bug fix patches. However, there are a lot of known instances where organizations have been slow to react and take precautions against security flaws. It is common for attackers to use recently-discovered vulnerabilities to attack an underlying system. If fixes are not quickly applied it allows attackers to exploit known loopholes and gain access to underlying systems and possibly even hosts.

7.5.2 SQL Injection Attacks

According to the Web Application Security Consortium (WASC), 9% of the total hacking incidents reported in the media before 27th July 2006 were due to SQL injection. More recent data from our own research shows that about 50% of the Web sites we have scanned this year are susceptible to SQL injection vulnerabilities.

Injection vulnerabilities are typically considered application-level threats and are quite often not considered during service design. Today many applications are being built as collections of services, or offered in the 'software as a service' model. However, in most cases these services are just designed as an additional layer on the existing application design. Hence the fallacies of application design simply permeate a level above to the services layer. In fact, as services are traditionally exposed through endpoints in human-readable formats, this just makes it easier for hackers to inject spurious requests to obtain unauthorized data. If a server does not validate data correctly, a SOAP message can easily be used to create XML data which inserts a parameter into an SQL query and has the server execute it with the rights of the Web service.

Let us explain this concept with a practical scenario that we came across during one of our security consulting engagements.

A global company had built its order management system and exposed a set of services to its business partners in order to route orders and locate inventory in real time. It also had a set of reporting Web services that allowed each business partner to view its own data. Let us consider a simple example of a Web service that allowed a dealer to view all its orders for a particular month. The following is a part of the sample Web service request, minus the requisite semantics for representation of the special characters:

```
<orderList>
 <dealer_code>  X&apoa; OR 1=1 --</dealer_code>
 <order_type>P</order_type>
```

In today's world of automated design and development, the following are the general next steps for execution of this Web service request:

(1) Parsing of the SOAP request.

(2) Binding the SOAP request to auto-generated objects, so an object Order Request with an attribute dealerCode.
(3) Selection of the orders for the dealer from the underlying table or tables as per the schema definition.

So it was represented as:

```
String sql =
"Select order_id, order_amt, ord_qty, order_status, part_number
From Order
Where dealer_code = "'+OrderRequest.getDealerCode()+'"
And dealer_status='C'
```

If it were replaced with the data from the service request above with direct binding, it would result in the SQL being represented to the database as:

```
Select order_id, order_amt, ord_qty, order_status, part_number
From Order
Where dealer_code = 'X' OR 1=1
--' And dealer_status='C'
```

The "--" is treated as comment by most of the database servers. This would lead to a particular dealer being able to view all the orders of his competitors.

It is also interesting to note that with WSDLs' and Web services' message parts, it has become even easier for hackers to make injection attacks. An interesting comment was made by a set of hackers who noted that most messages were direct representations of the underlying schema structures. For example, attributes like dealerCode were often represented in the underlying database schema as dealer_code. It has also been noted that most enterprises follow standards in schema definition, where database tables are named as 'tbl_Order' or 't_Order'. These standard representations are generally spread around among the hacker community and available to hackers in their own handbooks.

Hence it has been our experience that there is a higher risk of Web services being subjected to injection attacks, as compared to Web applications. The risk is heightened by the readily-available human-readable interface formats, which increase the probability of allowing hackers to the underlying data.

7.5.3 XPath and XQuery Injection Attacks

Most applications today use XML in different forms of interaction. With the increased adoption of new Web 2.0 platforms such as Ajax and RIA platforms such as FLEX there is even more XML moving over different transport protocols. There is also a strong federation of XML services from organizations such as

Google and Amazon, which rely heavily on the use of XML for everything from communication with backend services to persistence or even hosting your data with Web services.

A security designer needs to be aware of the threats and risks created by these approaches. One of the biggest threats is the possibility of XPath injection attacks. While the SQL injection attacks have been well publicized, the XPath injection attacks are still frequently not considered during security design.

However, XPath is a standard language, unlike the different versions of SQL used in various databases. Additionally, XPath gives complete access to the entire XML document, whereas access control can be restricted to users in an SQL database. Another, even more likely and possibly more troubling attack in XPath is the ability of attackers to exploit XPath to manipulate XML documents in an application on the fly.

The following is a sample of an XPath injection attack:

```xml
<?xml version="1.0" encoding="UTF-8"?>
<customers>
 <customer>
 <customerID>cust1</customerID>
 <firstname>fname1</firstname>
 <lastname>lname1</lastname>
 <SSN>1234567891</SSN>
 <status>A</status>
 </customer>
 <customer>
 <customerID>cust2</customerID>
 <firstname>fname2</firstname>
 <lastname>lname2</lastname>
 <SSN>1234567892</SSN>
 <status>C</status>
 </customer>
 <customer>
 <customerID>cust3</customerID>
 <firstname>fname3</firstname>
 <lastname>lname3</lastname>
 <SSN>1234567893</SSN>
 <status>A</status>
 </customer>
 </customers>
```

The following XPath statement can be used to get all the customer information, with the customer ID as cust1 and status as Active:

```
// customers / customer [customerID /text()='cust1' and
```

However, a hacker can easily modify the XPath query structure to represent it as:

```
//customers/customer[customerID/text()=" or 1=1 or "=" and
```

This will logically result in a query that always returns true and will always allow the attacker to gain access to the whole list of sensitive customer data. An attacker, upon spotting XPath-injection vulnerability in an XPath-based application, does not need to fully understand or guess the XPath query. Usually, within a few attempts, the attacker can generate a 'template' query data that can be used for a Blind XPath injection.

7.5.4 Blind XPath Injection

A more clinical approach to Blind XPath injection has been given in [4]. A couple of mechanisms are described below.

7.5.4.1 XPath Crawling

This method shows a technique to crawl an XPath document using scalar queries. The technique assumes no prior knowledge of the XML structure and works from an existing given XPath. The implementation very clearly demonstrates the ease with which the entire XML document can be reconstructed.

7.5.4.2 Query Booleanization

The uniqueness in this approach to XPath injection lies in the fact that it does not require data from the XML document embedded in the response, and that the whole XML document is eventually extracted, regardless of the format of the XPath query used by the application. This technique shows how a scalar XPath query (a query which returns String, Numeric or Boolean) can be transformed into a set of Boolean queries. This procedure has been described as 'Booleanization' of the query. Each Boolean query can be resolved by a single 'Blind' injection. It demonstrates the creation of an injection string including the Boolean query. When injected into an XPath query, it causes the application to behave in one way if the Boolean query resolves into 'true' and in another if the query resolves into 'false'.

7.5.5 Cross-Site Scripting Attacks

Cross-site scripting (XSS) attacks are extremely common with Web applications. The advent of Web 2.0 Rich Internet Applications has given a new lease of life to these attacks. The Web applications use XML over HTTP to retrieve content dynamically without submitting the entire content. This can be exploited by attackers to send malicious code, generally in the form of a browser side script, to a different end user. This generally happens when a Web application does not validate an input from a user and formulates the output which includes this input. This has increased because of the dynamic nature of Web 2.0 applications. UDDI references may be compromised through XSS, resulting in a reference to a malicious program instead of a valid Web service. Cross scripting attacks can also be orchestrated as a second step to WSDL phishing and session hijacking attacks. Attackers use XSS attacks to send malicious scripts and code to their victims, which gives them access to sensitive information. Possible results of XSS attacks are stated below:

(1) Encrypted SSL connections can become vulnerable.
(2) Persistent attacks can be launched, and even keystrokes can be monitored through malicious scripting.
(3) Domain-based security policies can be violated.

XSS attacks are generally sub-categorized into two forms, namely persistent code attacks and reflected code attacks.

7.5.5.1 Persistent Code Attacks

These attacks are mounted when malicious code is injected and persisted on the server. The victim gets infected with such an attack when he requests something from the server. These malicious code snippets are generally embedded in innocuous places like 'search the Website' or message forums. In the following example, one client embeds malicious HTML tags in a message intended for another.

```
This is the end of my message.
Hello message board. This is a message.
<SCRIPT>malicious code</SCRIPT>
```

7.5.5.2 Reflected Code Attacks

These attacks are orchestrated following a WSDL phishing attack. The victim is tricked into clicking on a malicious link. The attacker then injects malicious code, which comes back to the browser as if it is from a trusted source. Attackers frequently use a variety of methods to encode the malicious portion of the tag, such as Unicode, so the request is less suspicious-looking to the user. There are hundreds of variants of these attacks, including versions that do not even require any <> symbols. XSS attacks usually come in the form of embedded JavaScript. However, any embedded active content is a potential source of danger, including: ActiveX (OLE), VBscript, Shockwave and so on.

Sample XSS attacks [1]

We include typical XSS attacks below.

Simple check attack

The following string is a great way to check that there is no vulnerable JavaScript on a page.

```
";!--"<char:program_code><XSS></char:program_code>=&{()}
```

After injection, check for <XSS versus <XSS to see if it is vulnerable.

XSS locator

Injecting the string below results in the word 'XSS' popping up wherever a script is vulnerable, with no special XSS vector requirements. Another quick check is to inject the depreciated '<PLAINTEXT>' tag. If it is vulnerable it messes up the output.

```
';alert(String.fromCharCode(88,83,83))//\';alert
   (String.fromCharCode(88,83,83))//";alert(String.fromCharCode
   (88,83,83))//{";alert(String.fromCharCode(88,83,83))//-->
   </SCRIPT>">'><SCRIPT>alert(String.fromCharCode(88,83,83))
   </SCRIPT>
```

7.5.6 WSDL Probing

In Web services, the service interface is described in WSDL, to be exposed internally within the enterprise and externally to customers and business partners. WSDL contains meta-data information with advertising details including parameter details and API information. A WSDL file is a major source of information for an attacker. A WSDL description provides critical information like methods and input/output parameters. WSDL probing is one of the first things that a hacker does before going on to other attacks like parameter tampering and so on. It gives the hacker a lot of information, including the tools used to generate the WSDL files, underlying technologies and so on. In traditional attacks, this search for information would be followed by attempts to obtain entire IP address ranges. This attack is also called WSDL scanning or WSDL enumeration.

7.5.7 Enumerating Service from WSDL

The next step is to scrutinize the entire WSDL file and prepare a complete Web service profile.

The <service> element provides the name of the service and the entire access location for the same. This information gives the exact binding location for the client and the service to use with 'invoke'. It can be obtained from the following regex patterns: <service.*?> and <.*location.*[^>]>.

The WSDL <portType> element is a named set of abstract operations and messages. It indicates all the methods that can be invoked remotely. This is highly significant as these are the methods that can be targeted for initial sniffing.

WSDL refers to these port-type primitives as <operation>. The port-type name attribute provides a unique name along all port types defined within the enclosing WSDL document. These are the method names that are used with the 'invoke' operation.

Operations and messages are defined by a particular <portType> element. The message format and the protocol details of these operations and messages are defined by the WSDL <binding> element. There may be any number of bindings for a given <portType>. This allows a hacker to see the underlying transport protocols that can be used.

The WSDL <message> element consists of one or more logical parts, and each of these logical parts is associated with a type. These are used in conjunction with the <types> element. These represent the actual input and output messages and data types for a particular method. The <types> element uses XSD as the canonical type system for maximum interoperability and platform neutrality. This allows a hacker to check for methods and messages for known deficiencies like <any> elements being used.

This allows a complete Web service assessment profile to be created. On the basis of this WSDL assessment, the next step of detailing vulnerabilities and attacking them is carried out. In addition, the information provided in a WSDL file may allow an attacker to guess at other methods. For example, a service that offers stock quoting and trading services may advertise query methods like requestStockQuote, but also include an unpublished transactional method such as tradeStockQuote. It is simple for a persistent hacker to cycle through method string combinations (similar to cryptographic cipher unlocking) in order to discover unintentionally related or unpublished application programming interfaces.

7.5.8 Parameter-Based Attacks

Parameter tampering is a popular and simple attack targeting the application business logic. It is a popular attack for hacking into Web applications. Web servers are used to deliver the content. During a Web session, parameters are passed through the use of URL query strings, form fields and cookies.

A classic example of parameter tampering is changing parameters in form fields. The values presented in an HTML page can be manipulated by an attacker. In most cases this is as simple as saving the page, editing the HTML and reloading the page in the Web browser. Hidden fields are parameters invisible to the end user, normally used to provide status information to the Web application. Modifying a hidden field value will cause the Web application to change according to the new amount.

While the same problems do not exist for Web services directly, they manifest themselves in different forms. The proliferation of Web 2.0 technologies has led to Ajax-based interactions with Web services. Web services were designed for internal or B2B consumption, and therefore designers and developers often did not expect interaction with actual users. This lack of foresight led to some bad security assumptions during design. For example, the initial designers may have assumed that authentication, authorization and input validation would be performed by other middle-tier systems. Web services are now being invoked through asynchronous JavaScript calls. This leaves the Web service vulnerable to malicious JavaScript invocations. A real-life example of such usage is the consistent pitch from Microsoft to use Atlas hand-in-hand with Web services. Developers can now write JavaScript to create XML input and call the Web service right from within the client's browser. The XMLHttpRequest object is primarily used to invoke the Web services and can be done dynamically. Such an invocation can pass on the data displayed on the Web page. Anyone with basic knowledge of JavaScript can easily inject scripts on to the page and change the request object to send other data. If the method is GET, all form parameters and their values will appear in the query string of the next URL the user sees. An attacker may tamper with this query string.

Web services have traditionally been used for a singular request response communication paradigm. However, developers have designed their own session management mechanisms for managing transactions and client sessions. This has led to tokens or pseudo-cookies being passed across various Web service requests and is stored in hidden fields in Web pages.

Let us consider the following example.

The Amazon product catalogue Web service displays a set of products and their underlying prices. Once a user has placed the products they want in their shopping cart, it has to be linked to an external payment service. Let us assume the total amount is stored in a hidden field and sent to the payment service.

```
<input type="hidden" id="1976" name="cost" value="700.00">
```

If the application has not been designed properly, a hacker can save this page, and change the total amount field.

```
<input type="hidden" id="1976" name="cost" value="70.00">
```

The payment Web service can be invoked using XMLHttpRequest or by a traditional SOAP request call over XMLHttpRequest.

7.5.9 Authentication Attacks

Authentication attacks can take place at both request and response level.

7.5.9.1 Request Authentication Attacks

When an attacker gets possession of a legitimate client identity by theft or some other means it is called a request authentication attack. The attacker gets hold of legitimate credentials like passwords, digital certificates and so on. They can then make legitimate entries into a system, and the consequences can often be disastrous. The following is an interesting analogous real-life incident that happened in an enterprise.

When a user ID got locked or you forgot your password, you would have to call the help desk and let them know your employee number, and they would reset the password. In most enterprises there is no second-level check that is carried out. However, in this particular one a second-level verification existed and it was the user's birthday. A contractor who knew an employee's employee number and birthday called up the internal help desk and asked for the password to be reset. The security desk, having obtained the verification, gave the contractor 'fronting' as the employee the password. The employee in question was on an overseas trip for a week. The contractor now had legitimate credentials and was able to access and transfer large amounts of confidential data.

Forced-entry attacks

The primary reason most systems become vulnerable is that users tend to nominate weak passwords for their logins. Web services are just another interface layer over existing applications and are not treated any differently. However, unlike Web applications, XML Web service interfaces are heterogeneous in nature, with each underlying system having its own mechanisms for performing authentications. The most effective and simple technique is for the attacker to guess the password. This technique can be carried out either manually or via automated procedures.

Dictionary attacks

Dictionary attacks are where a hacker either manually or programmatically tries common passwords to gain entry into a system or multiple systems. Automated techniques allow hackers to programmatically try combinations of user names and passwords. Specialized custom tools like WebCracker and Brutus are readily available on the Internet. These tools attempt to gain access to a system by using predefined lists of usernames and passwords, taken from precomputed wordlists such as dictionaries.

Brute-force attacks

A brute-force attack is similar to the dictionary attack, except that it tries to break down cryptographic schemes. These could be digital certificates, tokens or even just the underlying private keys. The brute-force attack tries a large number of possibilities; for example, exhaustively working through all possible keys in order to decrypt a message.

Computed authentication attacks

These attacks are generally carried out when a specific algorithm generating a set of keys has been used. Previously, passwords and session IDs used to be generated in a nonrandom manner using mathematical algorithms. Once an attacker gets hold of a series of these randomly-generated numbers it is highly possible for him to judge the underlying technique used in generating the numbers. Most attackers also have a set of highly-advanced tools that have a built-in repository of the most often-used algorithms for ID generation. These tools take a series of generated IDs, evaluate them and respond with the algorithms that could have been used to generate them.

7.5.9.2 Stealth and Hijacking Attacks

These techniques involve stealing credentials or hijacking a user's session, which allows the attacker to get into an authenticated session. There are various techniques used to orchestrate a session hijack, such as:

Session fixation

A commonly accepted design technique when sending large data via Web services is to send a link that gives access to the data, instead of the data itself. In such scenarios, it is possible for an attacker to send a link to a user with a session ID they know, and then to impersonate the user once they log in. They are then free to cause damage.

Session hijacking

HTTP is a stateless protocol, and hence advanced session management techniques like cookies have been evolved for managing sessions in Web applications. However, session management for Web services needs to be designed and managed by the service providers. Session management techniques generally include some kind of session key to identify service consumers. An attacker can use sniffing techniques like packet sniffing to monitor and read network traffic between the parties, and steal a session key or cookie. This means that attackers can steal a session once the service consumer has been authenticated. Most Web services do not use encryption for the rest of the SSL operations once authenticated. This allows the attacker to impersonate the victim by continuing with the stolen sessions, even if the password itself is not compromised.

7.5.9.3 Response Authentication Attacks

A response authentication attack occurs when a malicious party poses as the service itself, fooling a client into making requests to it directly. Phishing and IP spoofing are known and commonly-propagated response authentication attacks.

WSDL phishing

Phishing is the term coined by hackers who imitate legitimate companies in e-mails to entice people to share passwords or credit-card numbers. Recently-reported cases have been with Best Buy and eBay, where people were directed to Web pages that looked nearly identical to the companies' sites. A Scandinavian bank was recently forced to shut down part of its Web banking service for 12 hours following a phishing attack that specifically targeted its paper-based onetime-password security system [5]. This has been extended to the world of Web services. WSDLs are exposed for public consumption. Attacks have been carried out by hosting manipulated WSDLs on fake URLs. The manipulated WSDL has all the same operations and bindings as the original WSDL. Generally one of the first operations the service consumer performs is to login, and when this happens the consumer inadvertently passes on their credentials to the attacker.

7.5.10 Man-in-the-Middle Attacks

Man-in-the-middle (MITM) is a form of eavesdropping in which the attacker makes independent connections with the victims and relays messages between them, making them believe that they are talking directly to each other over a private connection when in fact the entire conversation is controlled by the attacker. A phone wiretap is a prime example of eavesdropping, where a phone connection between two callers is tapped without the knowledge of either party.

The attacker can intercept all messages going between the two victims and inject new ones, which is straightforward in many circumstances. A MITM can only be successful if the attacker can hide themself from all involved parties for the length of the conversation.

By listening to the conversation between two systems, an attacker can collect useful data. XML being human-readable, it becomes even more easy to read and understand the data. Eavesdropping could be done simply to get access to data like credit-card account information, or it could be used to hijack identity credentials and start an authentication attack as described above. Eavesdropping also enables data integrity attacks, replay attacks and routing attacks.

7.5.10.1 Sniffing Attacks

Sniffing, or eavesdropping, is the act of monitoring traffic exchanged between two points in a network. This is one of the oldest forms of veiled attack carried out to capture sensitive information without the knowledge of the sender and the receiver. Web services can be used to capture sensitive plaintext data such as unencrypted passwords and security configuration information transmitted in SOAP, UDDI, WSDL and other such messages. With a simple packet sniffer, an attacker can easily read all plaintext traffic. Also, attackers can crack packets encrypted by lightweight algorithms and decipher payloads that the Web service developer considers to be secure. The sniffing of packets requires the attacker to insert a packet sniffer into the path between the service requestor and service provider.

7.5.10.2 Replay Attacks

A replay attack is a man-in-the-middle type of attack where a message is intercepted and replayed by an attacker to impersonate the original sender. A sniffer such as Ethereal [6] can capture traffic posted to a Web service. Testing tools like WebScarab [7] allow testers to resend packets to the target server. This allows attackers to intercept and resend the original message or change the message in order to compromise the host server. Digital signatures by themselves cannot prevent a replay attack because a signed message can be captured and resent. Replay attacks can be used by an attacker to create DoS attacks. These are the most difficult DoS attacks to control, as the attacker can flood a service with valid requests. Replay attacks are also known as spoofing attacks. Additionally, a *checksum spoofing* attack can be carried out by the attacker in between a requestor and provider. The message is signed with a hash to prove its integrity. The attacker intercepts the message and creates a new message with a recomputed hash using the original algorithm.

Data integrity attacks

The other and more deadly form of eavesdropping attack is known as a data integrity attack. The attacker eavesdrops into the traffic and modifies the data in the messages. Consider the example of a payment gateway Web service where users can transfer money between accounts or make an online payment. The attacker can change the target account number or hijack the financial details of the online payment.

7.5.11 SOAP Routing Attacks

SOAP messages travel from the service provider to the service consumer, potentially through a set of intermediaries. Intermediaries can process parts of a SOAP message as it travels from the origin to the destination. They are primarily used for:

(1) crossing trust domains
(2) scalability and high availability
(3) providing additional value-added services.

As shown in Figure 7.3, SOAP specifies a mechanism for targeting SOAP messages, but does not specify a mechanism for routing them. According to OASIS [8], various types of threat can target a SOAP message; for example, modification of SOAP message. The SOAP header that can be included in the SOAP message might be vulnerable. If an attacker obtains a SOAP message (for example, an implementation error) the data can be altered to modify or insert header instructions. According to a CERT Vulnerability Note (VU # 736923) [9], the Oracle Application Server 9iAS installed with SOAP allowed anonymous users to deploy and remove SOAP services. Consider that the attackers could create and deploy their own SQL statements. This could open the database behind the SOAP service for penetration and allow the attackers to insert a back door into the system [10]. WS-routing extends the SOAP model and provides a mechanism for routing SOAP messages between Web services. When messages are routed between Web services through intermediaries, they move across multiple SOAP servers, transforming content into multiple formats and so on. It is possible in between handovers from one intermediary to another for one intermediary to reveal information, deliberately or otherwise. The intermediaries can perform encryption and decryption. A compromised or malicious intermediary may participate in a MITM by adding false routing instructions and route the message to a malicious location. It may also be possible to forward on the document, after stripping out the malicious instructions and adding additional false instructions, to its original destination. The attacker can also redirect the messages to a nonexistent destination. This could lead to DoS attacks, as the messages will never reach their final destination and the underlying process can never be executed.

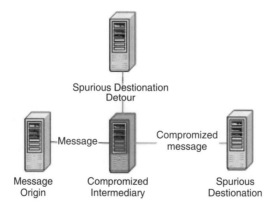

Figure 7.3   Compromised intermediaries via SOAP headers.

The WS-addressing specification provides a way to route XML traffic across different nodes in a service ecosystem. It operates by allowing an interim node in an XML path to assign routing instructions to an XML document. If one of these Web services way stations is compromised, it may participate in a man-in-the-middle attack by inserting bogus routing instructions to point a confidential document to a malicious location.

7.5.12 SOAP Attachments Virus

The Web service applications have business scenarios where they will have to deal with more than just XML data. There are instances when it is necessary to send binary data-like images and documents across the network. This has been addressed in SOAP with attachments. This is very similar to the evolution of emails, which now allow you to add attachments to your messages. However, just like the evolution of email attachments, where virus-based attachments could be sent to email addresses, there is another form of attack that can be orchestrated using SOAP with attachments. An attacker can send attachments which, when processed by the service, cause the service to malfunction and even compromise the underlying host.

7.5.13 XML Signature Redirection Attacks

XML signatures contain a Reference element that points to the signed data. The validation of the XML signature requires the target application to parse and include the content pointed to at the reference URI. Let us consider this example where the reference URI is:

```
<xdsig:Reference URI =
http://abc.com/download/Install/jdk1.6.exe>
```

The XML signature specification mandates the application to parse the URI to dereference it. So the XML processor must now download the large Java 6 executable and then compute the digital signature. This clogs up the network bandwidth and locks I/O threads till the file has been downloaded. Thus an XML signature dereference attack takes advantage of naïve Web service implementations and can cause extensive network clogging and bring down a service.

7.5.14 XML Attacks

XML has now become the de facto lingua franca for message interaction between applications. It is estimated that XML traffic on corporate networks has grown from 2% a few years back to around 50% today. It's used in security systems, enterprise applications, business process tooling and document management systems, and even modeling languages like UML use an underlying XML-based

XMI construct. However, XML documents, by their intrinsic nature, contribute to the lack of security in XML. The human-readable nature of XML makes it easy for attackers to decipher information and pick up the sensitive components like credit-card details or social-security numbers.

7.5.14.1   Coercive Parsing Attacks

Document-type definitions are still in use by some applications and services. These are still used for backward compatibility. The XML definitions allow for an element CDATA. The CDATA field allows the use of illegal characters. Attackers use the CDATA element feature to send possible system commands to the underlying systems. When querying a standard commercial XML parser, the CDATA component will be stripped, and the resulting string contains the nonescaped dangerous characters. The following example illustrates the attack:

```
<?xml version="1.0"?>
<?xml-stylesheet type="text/xsl" href="http://www.abc.com" ?>
<xsl:stylesheet xmlns:xsl="http://www.w3.org/TR/WDxsl">
<xsl:script>
<![CDATA[
x=new ActiveXObject("WScript.Shell");
x.Run("
*.dll");
]]>
</xsl:script>
</xsl:stylesheet>
```

It is evident that an attacker could sneak in system commands that could potentially be disastrous, as they could allow him to manipulate the host with a series of commands. They could also be used for injection attacks like XPath injection or XSS attacks.

7.5.14.2   XML External Entity (XXE) Attacks

These attacks are based on the ability of XML documents to reference data from outside the primary file. XML external entity attacks can be mounted on a system that parses XML from nontrusted sources. The XML processor may inadvertently open up HTTP connections or files. In the DTD, one declares the external reference with the following syntax:

```
<!ENTITY name SYSTEM "URI">
```

XML processor behavior is mandated by the specification as the following [11]:

'When an XML processor recognizes a reference to a parsed entity, in order to validate the document, the processor must include its replacement text. If the entity is external, and the processor is not attempting to validate the XML document, the processor may, but need not, include the entity's replacement text.'

However, if this external entity has been compromised by an attacker, the XML processor may implicitly parse data from this source. This could mean the processor is not 'VALIDATING' all the time and leaves itself open to vulnerabilities. A known such vulnerability was in the Peoplesoft application server, where people could mount XXE attacks and read any file under the Web server process. This meant that sensitive information could be exposed to attackers, and also opened TCP connections to the vulnerable servlet [12].

7.5.14.3   XML Bombs

XML bombs are message attacks that are created with the purpose of overwhelming the parser that processes these messages. This could result in the parser consuming too much memory, crash or DoS attack.

7.5.14.4   XML Denial-of-Service (XDoS) Attacks

XDoS attacks are orchestrated by attackers to make the response of services extremely slow or completely unavailable to legitimate users. This is generally accomplished by flooding the service with a large number of requests. These requests look like legitimate ones and force the service to process them. This leads to inordinate consumption of resources and in effect means that a legitimate request by a user cannot be serviced as its resources have been consumed by fake requests. The XDOS attacks can also be orchestrated by making complex XML requests like recursive payloads or jumbo payloads. These are described in more detail below.

Complex payloads

XML uses nesting to allow the creation of complex representation among elements. When an element is enclosed within another element it is termed as nesting. Nesting elements over three or four levels is a generally accepted practice for modeling relationships between entities. The parser generally consumes more resources as the nesting levels increase. An attacker might use the vulnerabilities presented in the schema to create complex payloads that require a large amount of resources to parse the underlying XML. The typical way is to create a recursive XML. While it depends upon the parser implementations, recursion does lead to a higher consumption of resources. The attacker typically composes a complex recursion of elements that crashes the parser. The other common techniques for causing such attacks are:

(1) Creating complex recursive nesting of elements.
(2) Using the <any> attribute and flooding the XML with a lot of elements.

(3) Creating a lot of attributes for an element.
(4) Opening and closing a tag too many times to create internal push and pop operations for the stack, as they are indicative of the start and end of elements.
(5) Creating an overflow attack by sending parameter data larger than the program can handle, causing stack and buffer overflow with the parser and the applications.

Oversized payloads

Attackers also use another tactic to create DoS attacks. Instead of sending a stream of messages or complex payloads, they just overwhelm the parser by sending a large message payload. The most common technique is to automate the creation of an <any> element that is defined as unbounded. This allows the attacker to create an unlimited number of elements under the unbounded tag. This creates a payload size which is extremely large (even in Gigabytes).

7.5.15 Schema-Based Attacks

Every XML is associated with its own schema. The schema is a template of the underlying XML document and can have a set of rules enforcing the validity of the XML document. This ensures that all XML schemas conform to a particular structure and allows the consumer of the XML document to parse the XML effectively. A schema file is what an XML parser uses to understand the XML's grammar and structure, and contains essential preprocessor instructions. There are various attacks that can be orchestrated using XML schemas, or against them.

Manipulating the schema can:

(1) Cause schema poisoning.
(2) Execute denial-of-service attack.
(3) Cause format changes (e.g. date and currency formats).

7.5.16 UDDI Registry Attacks

UDDI is an industry initiative to create a platform-independent, open framework for describing services, discovering businesses and integrating business services. It is designed as Yellow Pages, which allows businesses to publish their services. These services can be queried and searched by consumers. The registries typically allow service consumers to do the following:

(1) Search for enterprises by name.
(2) Search for services by specific categories.
(3) Search for support for specific protocols.
(4) Make location-based searches.

The result of a registry search is often technical information like the URL of a WSDL file, or business contact records. This allows the consumer to use the WSDL and interact with the service. The UDDI registries are themselves Web services, having full SOAP APIs and WSDL files. Thus the UDDI which contains a list of all services, with descriptions and WSDLs, is potentially the single largest source of information available to attackers. Attackers can typically use the registry as an information mine. This allows them to write automated scripts to query and search the UDDI and obtain the WSDLs. Once the WSDLs are obtained, attacks like WSDL scanning and probing can be orchestrated. The attacker can also hack the registry itself to modify the WSDLs or the underlying URLs. This will cause the service consumers to be redirected to fake service URLs when they search the registry.

7.5.16.1   Registry Disclosure Attacks

Attackers can use erroneously-configured registries (LDAP, X.500, etc.) to obtain critical information about the Web service being attacked. In particular, these registries can contain authentication information that an attacker may be able to use. Attackers can also use UDDI and ebXML registries to obtain information about the Web service being attacked. WSDL descriptions and audit logs are generally the most sought-after information.

Important points of information disclosure are the WSDL descriptions in the UDDI or ebXML registry, and the registry's audit logs. Further, these registries can be compromised or corrupted, which may allow an attacker to gain information about the Web service's host or even gain access to that host.

7.6   Services Threat Profile

Table 7.2 provides a view of the identified attacks and vulnerability groups, together with their mapping to services attack classification. The profile in Table 7.2 is a mapping of all the attacks to the various aspects of service interaction between a service provider and a service consumer. The mapping also includes the orchestration point for the attacks, which provides for a better ability to understand the attacks themselves. There is also an additional threat profile mapped to the National Institute of Standards and Technology (NIST) recommendations in their special edition 'Guide to Secure Web Services' [13].

Table 7.3 is another threat profile mapping of the attacks, categorized across various levels of security requirement as documented by NIST.

7.7   Conclusion

In this chapter, we have illustrated the core issues in service-level security. In particular, the higher level of loose coupling required for SOA mandates more

Table 7.2   Threat profile of service-level attacks.

Categories	Attacks	Attack orchestration point
Service attacks>	Known bug attacks SQL injection XPath and XQuery attacks Cross-site scripting attacks Parameter tampering Denial-of-service attacks	Service provider
Service communication attacks	Replay attacks Data integrity attacks XML signature redirection	Service communication layer (network and transport layers)
Service endpoint attacks	WSDL scanning WSDL phishing	WSDL interfaces
Service session attacks	Session hijacking Session fixation	Service session management
Service authentication attacks	Forced-entry attacks Brute-force attacks Dictionary attacks Computed authentication attacks	Service provider/UDDI registry attacks
Service protocol attacks	SOAP routing attacks SOAP attachment virus attacks	Service communication protocol
Service message attacks	External entity attacks Complex payload attacks Oversized payload attacks	XML messages communicated between provider and consumer
Service message template attacks	Schema poisoning attacks Coercive schema attacks	Message templates (XML schemas)
Service broker attacks	WSDL scanning WSDL phishing Parameter tampering Registry disclosure attacks	UDDI registry

flexible ways of handling security for SOA. Additionally, standards will be a key role as there is a need to interoperate across heterogeneous implementations of underlying systems. Finally, the openness and plaintext nature of XML-based distributed invocations are a cause of further complexity and higher vulnerability. We have explored in depth the various kinds of threat at service level, classifying them appropriately.

Table 7.3 NIST standard of service-level attacks.

NIST classification	Attacks
Reconnaissance attacks	WSDL scanning
	WSDL phishing
	Directory traversal attack
Confidentiality attacks	Sniffing
	XML signature redirection
Integrity attacks	Parameter tampering
	Schema poisoning
	External entity attack
	SOAP routing attacks
	Spoofing
	Data integrity attacks
	Replay attacks
Denial-of-service attacks	Complex payload attacks
	Oversized payload attacks
	Schema poisoning
Command injection attacks	SQL injection
	Xpath injection
	Cross-site scripting
Malicious code attacks	Command injection
	SOAP attachment virus attacks
	XML bombs
Privilege escalation attacks	Schema format attacks
	Buffer overflow attacks

References

[1] Padmanabhuni, S. and Adarkar, H. (2005) Security in service oriented architecture: issues, standards and implementations, *Service oriented Software System Engineering, Challenges and Practices*, IGI Global.

[2] UDDI version 3.0.2 is available at http://www.oasis-open.org/committees/uddi-spec/doc/spec/v3/uddi-v3.0.2-20041019.htm.

[3] NIST Special Publication (SP) NIST 800-95, *Guide to Secure Web Services*.

[4] Klein, A. *Blind XPath Injection Attack*, http://www.packetstormsecurity.org/papers/bypass/Blind_XPath_Injection_20040518.pdf.

[5] http://www.finextra.com/fullstory.asp?id = 14384

[6] http://www.ethereal.com

[7] http://www.owasp.org/index.php/Category:OWASP_WebScarab_Project

[8] OASIS (2004). *Web Services Security: SOAP Message Security 1.0*. White paper, 2004.

[9] http://www.docs.oasis-open.org/wss/2004/01/oasis-200401-wsssoap-message-security-1.0

[10] Vulnerability Note VU#736923 http://www.kb.cert.org/vuls/id/736923

[11] Clewlow C, SOAP and Security, accessible http://www.qinetiq.com/home/security/digital_security/white_paper_index.Par.0013.File.pdf

[12] Clewlow, C. (2004). *SOAP and Security*. White paper, 2002. Date: December 2004.
[13] Moore et al. (2001). *USENIX Security*.

Further Reading

Apache Axis, http://ws.apache.org/axis/.

Apache Web server, http://httpd.apache.org/.

Boubez, T. (2004) *The Challenges of Web Services Security. 'Beyond XML Firewalls' Security Coordination for Web Services*. August 3, 2004.

Demchenko, Y. (2004) *'White Collar' Attacks on Web Services and Grids. Grid Security Threats Analysis and Grid Security Incident Data Model Definition*. August 2004.

Demchenko, Y., Gommans, L., de Laat, C. and Oudenaarde, B. (2005) *Web Services and Grid Security Vulnerabilities and Threats Analysis and Model*. The 6th IEEE/ACM International Workshop on Grid Computing, pp. 13–14.

Derry, J. *Advanced Web Services Security & Hacking*. http://www.owasp.org/images/3/3a/OWASP AppSec2006Seattle_Web_Services_Security.ppt

Extensible Markup Language (XML) 1.0W3C Recommendation, http://www.w3.org/TR/REC-xml# include-if-valid. August 16, 2004.

http://www.sarvega.com/xml-security-products.php

http://www.xforce.iss.net/xforce/alerts/id/advise139

http://www.ibm.com/developerworks/xml/library/x-xpathinjection.html

http://www.acunetix.com/websitesecurity/authentication.htm

http://www.xwss.org/Malicious_Attack_Protection_for_XML_Web_Services.html

http://www.actional.com/resources/whitepapers/Web-Service-Risks/Web-Services-Hacking.html

http://www.reactivity.com/products/

Moradian, E. and Håkansson, A. (2006) Possible attacks on XML web services. *International Journal of Computer Science and Network Security*, **6** (1B), http://www.paper.ijcsns.org/07_book/ 200601/200601B48.pdf

O'Neill, M. XML and Web Services: Are We Secure Yet.

Shah, S. *Web Services: Enumeration and Profiling*. http://www.net-square.com/whitepapers/ WebServices_Profiling.pdf

Stamos, A. (2005) *Attacking Web Services*, OWASP AppSec DC http://www.owasp.org/images/d/d1/ AppSec2005DC-Alex_Stamos-Attacking_Web_Services.ppt

Web Services Security: Non-Repudiation. Proposal Draft 05, April 11, 2003.

XSS Attack examples http://www.ha.ckers.org/xss.html

XSS Cert advisory http://www.cert.org/advisories/CA-2000-02.html

8

Host-Level Solutions

8.1 Background

In Chapter 4 we covered various vulnerabilities and threats that might affect a host. The issue with transient code that runs on hosts is that it cannot be trusted and can be malicious. Running it directly on a host could cause severe damage to the system. Similarly, genuine transient code is at the mercy of executing hosts and is very vulnerable to attacks from malicious host applications and users. Resident code, though trusted, historically has contained bugs that have made systems vulnerable. A simple yet effective solution to both of these issues lies in creating 'isolated' environments for each, in such a way that the ill-effects they cause are contained within their own environments and cannot affect the system as a whole. Sandboxing and virtualization [1] are two key isolation techniques that we will study in this chapter. They are useful in addressing the privacy and security issues discussed in Chapter 4.

The other way to guarantee host security in distributed environments is to execute 'trusted' code and disallow everything else. To make this possible, every piece of mobile code must carry along with it a tamper-resistant proof that it is trustworthy. Proof-carrying code (PCC) is one technique to certify a mobile code as trustworthy.

In relation to resource starvation issues, we will look at techniques such as reservation, priority and resource management. These approaches are useful in ensuring a fair share of resources among competing applications/jobs.

We will also look at solutions, such as program shepherding, to protect hosts against code-injection attacks.

8.2 Sandboxing

Sandboxing is a very popular technique for achieving isolation. The key operating principle behind this technique is that an application can cause very little

Distributed Systems Security   A. Belapurkar, A. Chakrabarti, S. Padmanabhuni, H. Ponnapalli, N. Varadarajan and S. Sundarrajan © 2009 John Wiley & Sons, Ltd

harm if access to resources (through the operating system (OS)) is controlled or restricted. System calls are the only means in modern systems for users to gain access to resources. By restricting system call access, it is possible to contain the ill-effects of any malicious code executing on the host. There are four ways to create an isolated sandbox in which to execute untrusted code, namely kernel-level sandboxing, user-level sandboxing, delegation-based sandboxing and file-system isolation. We will look at each of these approaches, with examples.

8.2.1 Kernel-Level Sandboxing

The simplest way to create a sandbox is to hook on the 'system call entry table' and redirect the calls to a kernel-loadable module [2]. The kernel module then has the ability to inspect every single system call made by the user processes. The decision to allow or deny access to a system call is made by a policy engine operating in either the kernel space or the user space. The policy engine can be configured to allow or disallow operations based on accessed resources, users and processes. Figure 8.1 depicts a typical kernel-module-based sandbox.

Kernel modes Janus, Systrace and Remus all fall under this category. System calls that require special handling to provide the sandboxing facility are intercepted and handled by the hooked kernel module in consultation with the policy engine, which typically runs at the user space. Any innocuous function calls such as close, exit and so on are directly executed by the kernel and need not be intercepted. The policy engine is involved in all cases where system calls are intercepted. Depending upon which user/process is invoking the system call and the argument

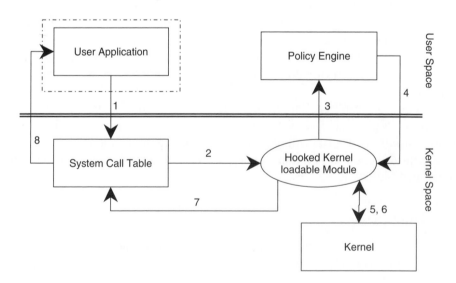

Figure 8.1   A kernel-module-based sandbox.

associated with the call, the policy engine either returns an allow or a deny. If a system call is allowed, the kernel sandbox module directly makes the actual system call, whose entry pointers have been preserved during the initialization phase.

This doesn't require any change to the application binary or the way a process is invoked. Once the system call hooks are placed, all processes in the system will have to go through this, with no exception. This can introduce significant performance overhead, particularly for systems which rely heavily on system calls.

8.2.2 User-Level Sandboxing

User-level sandboxing is different from the kernel-loadable module technique in that it intercepts the system calls not through a hook on the system call table, but by tracing a process using standard debugging/trace features provided/supported by the OS. For example, in UNIX and similar systems, the ptrace function allows a parent process to trace any of its child processes. This tracing ability can be used to identify the system calls invoked by a user application, along with the arguments used during invocation. The use of a policy engine to determine whether a particular system call is to be allowed or denied is similar to the kernel-loadable module technique. Figure 8.2 shows how this user-level sandboxing is implemented.

The user-level sandboxing version of Janus [3] adopts a similar approach to that depicted above. The application which requires sandboxing is run under a parent process, which enables tracing of its permissions. Through the tracing, all system calls that the application makes can be trapped. This information, along

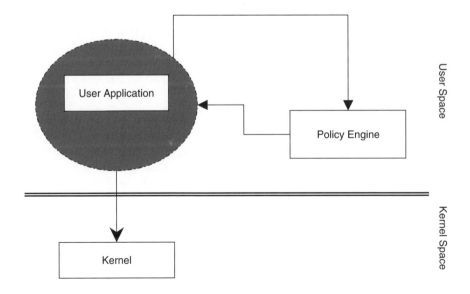

Figure 8.2 User-level sandboxing.

with the system call arguments, is sent to the policy engine. The policy engine can then deny or allow the operation. Unlike the kernel-mode system-call interception technique discussed earlier, all the system calls in this technique will be trapped by the tracing parent process. This approach typically has higher overhead than a kernel-based solution. Also, the application which requires isolation will need to run under a parent sandboxing process.

8.2.3 Delegation-Based Sandboxing

Kernel-level and user-level sandboxing techniques suffer from race conditions [4] due to nonatomicity of policy evaluation and system-call access. Delegated sandboxing architecture (Figure 8.3) can alleviate the race condition issues. In this approach, when a user process invokes a system call the system makes a callback to an emulation library in the user space, which provides isolation service. The callback function in the emulation library calls a user-space delegation agent, which actually executes the system call on behalf of the user application.

Ostia [5], developed at Stanford, is a good example of such a sandbox solution. The key aspect to note here is that the delegation agent acts on behalf of the user process to invoke the system call, and there is a separate delegation agent process for each user process created through this system. The key advantage, besides solving the race condition [6] (not all race conditions are addressed), is that the kernel module is very minimal and can be ported easily, as dependency on the kernel structures is minimal.

8.2.4 File-System Isolation

This approach attempts to completely isolate file-system changes from the main or primary file system. To start with, the solution [7] uses system-call interception to hook into all system calls that can alter the file system. The hook implemented with

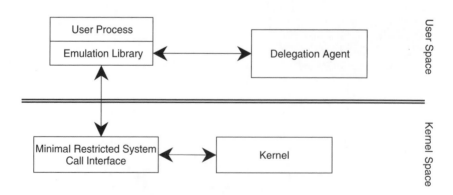

Figure 8.3 Delegated sandboxing.

the system creates an isolated shadow file system, which records all file-system changes while leaving the primary file system untouched. In other words, when a file is updated, the solution uses copy-on-write to duplicate the file and maintain a separate copy, leaving the original as it was. All child processes will use the newer, isolated version of the file and not the original copy. This technique can be used to completely isolate changes to a file system by untrusted code. Upon program completion, the shadow file system becomes available to an administrator or other system user for review. Once they are convinced that the application is trustworthy, it can be run outside the sandbox.

8.3   Virtualization

System virtualization [8–10] is not a recent innovation, but has been around for over three decades, particularly in mainframe environments. Virtualization on commodity hardware was not traditionally attractive. However, significant increase in computing capability on commodity hardware (based on x86 architecture) has renewed interest in this area. In this second wave of adoption, use of virtualization goes beyond system partitioning. The properties of virtualization, namely resource control and isolation, make it appealing to use in solving host-level security issues [11]. Before we look at how virtualization can provide solutions to the various host security threats discussed in Chapter 4, let us look at different types of virtualization and how they are implemented. The formal requirements [12] for virtualizable system architecture are:

(1) *Efficiency:* ability to execute innocuous instructions directly on the hardware.
(2) *Resource control:* all system-resource access should be routed through a virtual machine monitor (VMM).
(3) *Equivalence:* application behavior should be the same as for a nonvirtualized platform.

Based on how the platform is virtualized, system virtualization can be broadly grouped into four categories, namely full-system virtualization, para virtualization, shared-kernel virtualization and hosted virtualization.

8.3.1   Full-System Virtualization

In this type of virtualization, the VMM sits directly on top of the hardware, abstracts it, and allows creation of a virtual machine with properties identical to it. The VMM handles all resource access on the host, and none of the virtual machines (VM) has direct access to any physical resources. Instructions, other than the ones used for resource access, execute directly on the physical hardware, thus reducing any CPU virtualization overheads. Z/OS is a good example of full-system virtualization. On the x86 environment, VMWare ESX 2.5 can be classified as a full-system virtualization VMM.

VMWare ESX 2.5 [13] virtualizes the x86 platform using a combination of direct execution and instruction virtualization to support unmodified guest OSs. VMWare ESX virtualizes all the system resources such as CPU, memory, disk and network. CPU virtualization is achieved by directly executing instructions when they are mostly unprivileged code (such as those contained inside user-mode applications/processes) and virtualizing privileged ones (such as those relating to the OS). The VMM emulates standard devices, while the hardware interface layer in VMWare has custom drivers for specific devices. The device drivers in the hardware interface layer are involved in actually accessing the physical resources. The emulated devices simply create a communication channel between the guest OS device drivers and the drivers in the hardware interface layer. The VMWare kernel is responsible for mapping the physical pages of the virtual machine to the actual machine pages on the physical host, thus virtualizing memory access. Figure 8.4 describes the architecture of this solution.

8.3.2 Para Virtualization

Para virtualization [14] is a technique through which an instruction set of the hardware which does not support virtualization is modified to a fully-virtualizable instruction set, such that the system can be fully virtualized. Modifying the instruction set of the hardware means that the OS (read kernel) will require porting to the newer instruction set. In most cases applications do not require porting, because they do not use instructions that operate with the hardware's internal registers or machine state.

By creating a parallel virtualizable instruction set through para virtualization, a hypervisor can be built which can host the kernels of the guest OS. No run-time translation will be required in this case as the kernel porting is done at compile time, and the performance in most cases is similar to that experienced in full-system virtualization.

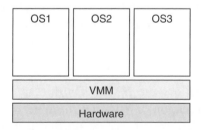

Figure 8.4 Full-system virtualization.

The biggest negative with this technique has to do with the fact that the kernel needs porting for the hypervisor. This may not be possible with a proprietary OS/kernel such as Windows.

Native device drivers are installed in one of the guest OS instances and typically the other guest OSs accesses them through device channels, as in mainframes. In other cases the hypervisor may include device drivers for a limited number of devices and the guest Oss access the devices through the hypervisor. In both cases the number of context switches between the guest OS and VMM are limited, resulting in a near-native I/O bandwidth. Xen is the most popular para virtualization solution on the x86 platform.

Xen [15, 16] is an open-source para virtualization solution conceived by researchers from the University of Cambridge. The heart of Xen constitutes the VMM, which is also termed the hypervisor. This hypervisor exposes itself as a platform for multiple VM on which guest OSs run. It also abstracts the hardware access. The guest OSs need to understand the hypervisor's hardware abstraction layer in order to access the underlying hardware, hence their kernels are modified accordingly. As in the case with VMWare, Xen directly executes the guest OS application instructions on the hardware and the precompiled guest OS code makes calls to the Xen hypervisor to negotiate resource access. Xen maintains software page tables to manage guest OS physical and virtual memory. Emulated standard devices in the administrator domain, where requests and responses by/for each guest OS are placed in a simple ring structure, manage I/O access. Shared memory is used to exchange this information. See Figure 8.5.

8.3.3 Shared-Kernel Virtualization

Shared-kernel kirtualization [17], as the name suggests, is not based on system virtualization, but rather on OS virtualization. It is a way to partition a single large machine into several virtual private machines. Each virtual private machine shares

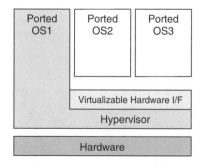

Figure 8.5 Para virtualization.

the same kernel, is isolated from the others and has fine resource commitment guarantees. The approach is to share a single kernel among multiple containers (partitions) by creating a context within which processes run and the file system is maintained.

Each container belongs to a context that is assigned to it when it is created, and all access to system resources is validated against the context. If a container requests access to a particular resource not owned by its own context, the container will be refused access to it by the kernel. Using the context, the kernel will also be able to share the resource equitably, based on some policy between the various containers. As is apparent from Figure 8.6, there are no additional levels of indirection (as in the case of a hypervisor) and the performance is very close to the native performance. All features available on the bare metal machine are also available to each virtual partition.

Since the kernel is shared, the isolation is not as robust as the other virtualization architectures. In addition, the host kernel needs to be modified to be made aware of the context, which is otherwise not implemented in the kernel. Solaris Zones [18], Virtuozzo [19], OpenVZ and Linux V Server [20] are some well-known implementations of shared-kernel virtualization.

Solaris Zones [21] is widely used for consolidation and other isolation purposes on Solaris 10 or higher platforms. It has a notion of a global zone (which is the administrator zone) and several nonglobal zones. The zone-aware Solaris kernel creates a separate namespace for each of the nonglobal zones and ensures that resources created in a nonglobal zone are not available outside it. The global zone is an exception. The *'root'* user within the global zone has certain privileges to monitor and manage resources that belong to other, nonglobal zones. Solaris uses a technique similar to BSD jail to create restricted file-system access, including the /proc file system. Solaris creates logical network interfaces in the global zone, as in any multihomed server, and maps these logical interfaces to each nonglobal zone as independent network interfaces. This gives distinct network identity to each of the nonglobal zones.

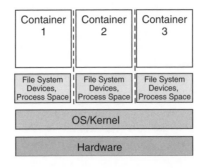

Figure 8.6   Shared-kernel virtualization.

8.3.4 Hosted Virtualization

This is a technique [22, 23] used very specifically on x86 environments to address their nonvirtualizable instruction set. In this model, the VMM is actually an add-on to an existing OS, which exists as a host, with all the VMs as guests. The guest OSs, unlike in the other models, run at the same privilege level on the hardware as the application. The major difference between this and the others is that the hosted virtualization uses the device drivers in the host OS for I/O virtualization. CPU and memory are similar to how they were in the case of VMWare ESX mentioned earlier. On-the-fly translators translate privileged guest OS code, while application code executes directly on the CPU. Figure 8.7 shows a schematic representation of how a hosted model works. The I/O VMM actually runs on top of the host OS. All I/O resource access by the guest OS requires intervention of the host OS. VMWare Server, VMWare Workstation, MS Virtual PC and MS Virtual Server [24] are all examples of the hosted virtualization model.

VMWare Server has two main components, namely the VMDriver, which acts a VMM, and the VMApp, which provides I/O virtualization through the host OS. Instructions inside guest OS applications are executed directly by the hardware, while guest OS instructions are virtualized using binary translators. Any device-access request by the guest OS is sent to the VMApp running inside the host OS. The VMApp leverages the native device drivers already set up on the host OS to handle the device-access requests.

8.3.5 Hardware Assists

In a typical IA32 architecture processor (Figure 8.8) such as an Intel Xeon processor, there are four privilege rings. The lowest ring level, 0, is where the OS runs, and ring level 3 is where the user applications run. Ring levels 1 and 2 are reserved or not used.

In virtualized platforms, VMM needs to be in absolute control of the system resources. This requires VMM to run at ring level 0 and the guest OS to run at a ring level higher than 0. This is termed ring deprivileging (Figure 8.9).

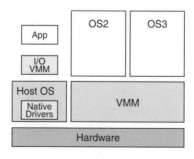

Figure 8.7 Hosted virtualization.

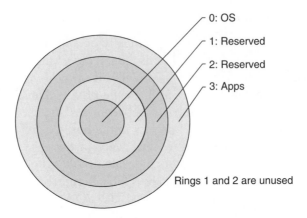

Figure 8.8 IA32 architecture.

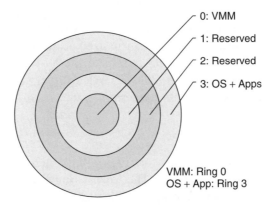

Figure 8.9 Ring deprivileging.

Ring deprivileging introduces a number of virtualization issues.:

(1) *Ring aliasing:* this is a problem when software is run at a ring level different to the one at which it was expected to run.
(2) *Address-space compression:* VMM requires a certain address area reserved for itself, which will impact the linear address space available to any process inside a guest OS.
(3) *Nonfaulting access to privileged state:* instructions that are expected to be used only in ring level 0 do not result in a fault when executed at a higher ring.
(4) *Ring compression:* running both guest OSs and the applications on these guest OSs on the same ring level means that the guest OSs are not protected from the guest applications.

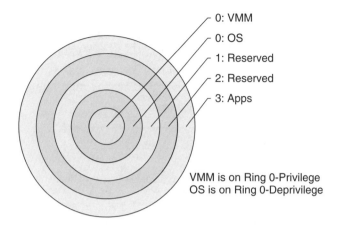

Figure 8.10   Additional VMM level.

In all there are around 17 instructions in the IA32 architecture that are not virtualizable. This issue has been solved with the introduction of VT (virtualization technology) in the latest generation of the Intel processor [25, 26]. The new processor introduces an additional VMM ring level to address the ring deprivileging issue (see Figure 8.10). VMM now runs as VMX root operation, while the guest OSs run as VMX nonroot operations. A transition from VMX root to nonroot and back is achieved through new instructions, namely VM entry and VM exit respectively. In VMX nonroot operation, instructions that requires VMM intervention cause unconditional VM exits. This makes it possible for unmodified guest OSs to run directly on the hardware with minimal performance overheads. VMMs that run on VT do not require any binary translation support for virtualizing guest OS privileged code.

8.3.6   Security Using Virtualization

Isolation is a fundamental property of virtualization. A virtualized environment ensures that each partition is completely isolated from other partitions. Apart from fault isolation, virtualization can provide resource isolation by managing resource allocation among various partitions. Virtualized execution environments can address issues [27] relating to eavesdropping, job faults and resource starvation. Figure 8.11 gives a simple depiction of how virtualization can help in addressing privacy and security concerns. Each partition/compartment has an independent, secure view of its data, processes, devices and memory, which is not accessible by the other partitions. A resource manager running alongside the VMM can also ensure that no single partition runs away with all the system resources, and the resource apportioning is strictly based on policies established for this purpose.

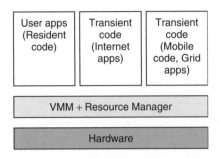

Figure 8.11   Resource manager and isolation.

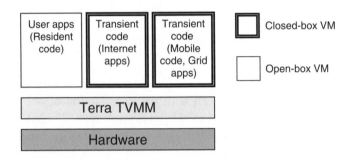

Figure 8.12   Terra architecture.

Terra [28], proposed by Garfinkel *et al.*, illustrates how a trusted computing platform can be built over a virtualized environment. Terra system architecture is depicted in Figure 8.12.

Terra has a tamper-resistant trusted VMM that can host commodity or specialized thin VM. The VM can be in either a closed-box mode or an open-box mode. The open-box VM can run commodity OSs and regular software stacks, while the closed-box VM is a specific-purpose appliance. The TVMM can prove that the closed-box VMs have not been tampered with to remote clients which require services from them, and the attestation process allows remote VM owners to verify whether or not a closed-box VM has been tampered with. A certificate chain for each of the components in a closed-box VM, from the boot loader right up to the software stack, is used for attestation purposes.

Besides virtualizing hardware, a VMM can play a vital role in resource management. System resources such as CPU, memory and other devices can be shared based on policies or using the resource manager that is normally bundled with a VMM. For instance, it is possible to allocate the CPU to various VMs based on shares allocated to each VM, or on virtual CPUs assigned to them. A VM with two virtual CPUs would get scheduled on the physical CPU twice as often as a VM which has been assigned only one. It is also possible to attach priorities to VMs, in a similar way to process priorities. Such resource-management capabilities are very handy when it comes to addressing the resource-starvation issues detailed in Chapter 4.

8.3.7 Future Security Trends Based on Virtualization

Security software stacks, such as intrusion detection systems, intrusion prevention systems, firewalls and so on, have traditionally run directly on physical hosts. Smart malwares have looked to disable these software agents before spreading through a network, in order to avoid being noticed and increase damage manifold. In future, such security stacks will be run inside a separate security VM [29]. This will ensure that when an application/VM is compromised, the security VM is out of reach and can continue to detect and prevent security attacks.

8.3.8 Application Streaming

Application virtualization and streaming products such as Sun Global Secure Desktop [30] and MS SoftGrid [31], and virtual desktop infrastructure (VDI) products such as VMWare VDI [32], Sun VDI [33] and so on are gaining rapid ground. These solutions make it possible for applications to be hosted on a central data centre, while streaming just the user interface to users' desktops/laptops. This will mean that desktops no longer have data or software installed locally. In a way, the desktops will be mere devices through which users interact with their applications, while the applications themselves are running elsewhere.

By moving the application execution and confidential data away from the desktop/host, this technology significantly limits the damage transient code can cause to them. While application virtualization and VDI by themselves cannot solve the security issues related to transient code, when used alongside other technologies they can prove very useful in thwarting host security threats.

8.4 Resource Management

In a shared computing facility such as a grid or cluster, managing resources for different job/application execution requests is crucial in order to avoid issues relating to resource starvation. Techniques such as advance reservation, priority reduction and so on could be used to some extent to address this, and resource allocations can be precisely managed through use of workload management solutions such as Solaris Resource Manager (SRM), Windows System Resource Manager (WSRM), Citrix ARMTech and Entitlement-based scheduling (EBS).

8.4.1 Advance Reservation

In this technique [34], a cluster/grid host explicitly reserves resources (CPU, memory, etc.) for executing a job. For instance, a dual CPU host can set aside a processor for grid jobs, while keeping a processor for the native host jobs. Similarly, memory can be set aside through system calls which limit the physical memory (resident set size) available for a grid job. CPU time can also be limited by using appropriate system calls. These controls, when used in tandem, can reduce issues relating to resource starvation on the host.

8.4.2 Priority Reduction

To ensure that native jobs do not starve, the host must process priorities effectively. Priorities for grid jobs can be lowered [35] such that native jobs continue to perform without any impact. However, this is very conservative and can result in the creation of a cycle-stealing environment. Other priority-management schemes such as RRDP, used in the Sun Grid Engine, can be more effective in handling resource-starvation issues. RRDP uses a more holistic priority-reduction technique based on process-wait time and deadlines, besides other factors.

8.4.3 Solaris Resource Manager

In general, process priorities decide when and for how long a particular process is scheduled to run on the processor. They do not provide adequate control to manage the resources allocated to a process precisely. SRM [36] provides a way to manage the resources (CPU, memory, file, thread and process limits) available to a workload. Workload can be defined as a group of related processes tackling a common problem. SRM has notions of projects and tasks, which are logical groupings of processes owned by a user or a group, to which relative or discrete resources can be assigned.

An SRM project can be allocated a certain number of CPU shares. The relative share allocation among various projects determines how large a percentage of the CPU is apportioned to each. For instance, if two groups, namely 'default' and 'user.root', are allocated 80 and 20 shares respectively, the processes running under 'default' tend to consume 80% of CPU time and processes under 'user.root' consume roughly 20%. Individual processes within these projects are scheduled based on their relative priorities. The fair-shares scheduler within the Solaris kernel has to be enabled for SRM to enforce these resource controls.

The resource cap daemon in SRM can similarly manage the resident set size (RSS) (physical memory) available to a project group. The memory cap for each project can be set, and SRM will ensure that the physical memory consumption by the project does not exceed the cap.

Similarly, other resources such as threads, file handles and so on can be capped for each project. These resource controls can be used to manage grid and native host jobs without causing any resource starvation. SRM currently does not have capabilities to handle network bandwidths or disk storages. It does, however, have extended accounting support to precisely measure resources consumed by various tasks and projects.

8.4.4 Windows System Resource Manager

WSRM [37] is a recent introduction by Microsoft, designed to manage resources on the Windows 2003 server family. Through WSRM, it is possible to create resource-allocation policies either by user or by process. CPU and memory resources can be controlled granularly using WSRM.

Soft CPU caps on processes or users are set through policies based on percentage CPU allocation. WSRM manages the individual process priorities and processor affinities to meet the soft caps set up in the system. It is also possible to limit the physical memory (working set size) and the paging limit for an individual process or for all processes running under a user.

WSRM can be used in conjunction with the grid/cluster scheduler to ensure that native applications are not starved for resources. However, WSRM use is limited to the Windows Server 2003 family.

8.4.5 Citrix ARMTech

Aurema's ARMTech [38] provides active resource-management capabilities for the Windows server family. It shares lot of similarities with SRM. CPU or memory resource allocation can be done based on users, applications or application groups.

ARMTech allows administrators to group specific processes running on the host, to form application groups. CPU percentages or shares are associated with each of the resource consumer groups (users, applications or application groups). The amount of resources allocated to a consumer resource group is relative to the number of shares assigned to it.

ARMTech allows administrators to extend the resource control to a hierarchy of resource consumer groups using a tiered policy path. Resource-allocation policies can be specified at more than one tier or layer. ARMTech publishes accounting information on resource usage through the standard WMI interface available on Windows server platforms.

8.4.6 Entitlement-Based Scheduling

EBS [39] is a scheduler for Linux OS that can be used to provide similar resource management capabilities to SRM or ARMTech. Conceptually, EBS is very similar to other resource managers and uses the same notion of shares or relative entitlements for processes. Once created, a process's entitlement can be set, and the Linux scheduler will use this to ensure that the process gets its relative share. In Linux, the RSS of a process can be directly managed through system calls. These system calls, together with EBS, can provide the necessary resource-control capability to address any resource-starvation issue on the Linux platform. EBS is a separate add-on to the Linux kernel and is not merged into the main kernel source.

8.5   Proof-Carrying Code

The reason security and privacy are very serious issues in distributed systems is that the mobile/transient code that is downloaded and executed on physical hosts is not trustworthy. All the techniques discussed so far relating to sandboxing and virtualization are aimed at providing the necessary isolation at the host level to protect the data and system from malicious users or faulty mobile code. While these techniques are widely used today to provide the necessary security cover, there are perhaps other ways to address host-level security and privacy. One method is to leave the responsibility of proving the trustworthiness to the mobile/transient code. In other words, it is the responsibility of the transient code author to show that the code is safe. The execution host simply verifies that the claim by the author is true and confirms that the code has not been tampered with. Once satisfied on both counts, the execution host can safely execute the mobile code.

Necula and Lee originally proposed the idea of mobile code carrying proof of its safety, calling it PCC [40]. There are two parties in this system, namely a *code producer* and a *code consumer*. The code consumer defines a *safety policy* and the code producer demonstrates conformance to this safety policy. The safety policy consists of *safety rules* and *interfaces*. The safety rules are a list of permissible operations and preconditions which are to be satisfied before the operations are performed, while the interfaces are contracts that the code should conform to. The safety policy is published by the code consumer and is readily accessible to all code producers. Once established, the safety policy can be used by the code producers to write code that conforms to the consumer's wishes, and to provide proof of its conformance.

The three life stages of a PCC are *certification*, *validation* and *execution* (see Figure 8.13). In the certification stage, the PCC system computes the safety predicate of the program and generates a proof of that predicate. The responsibility of validation lies with the code consumer. The proof of the safety predicate is in a binary format embedded within the executable; the consumer checks whether the safety predicate and proof match with the actual code. It is adequate to perform this validation once before executing the code repeatedly on the host.

Once validated, there are no run-time overheads with this solution, unlike in isolation techniques such as sandboxing or virtualization. However, in a distributed

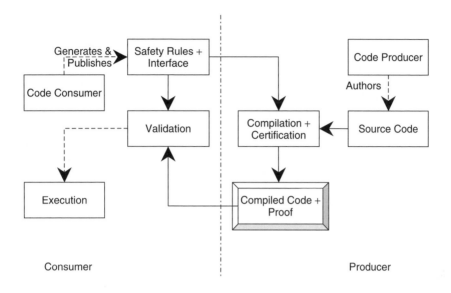

Figure 8.13 Proof-carrying code lifecycle.

computing environment there may be many consumers (potentially in the millions if the code is Internet-based) and it is not possible to include the safety policies of every single consumer.

8.6 Memory Firewall

The techniques we have discussed thus far address security threats and vulnerabilities relating mainly to transient or mobile code. The effect of transient or mobile code vulnerability is mostly confined to desktops or workstations. However, most host security issues on business-critical servers arise from vulnerabilities that exist in the resident or trusted code in the form of buffer overflows. Buffer overflow problems, as discussed in Chapter 4, predominantly occur due to use of unprotected buffer/string-handling functions.

Kiriansky *et al.* from MIT have proposed a technique known as *program shepherding* [41] (also widely known as *memory firewall*; see Figure 8.14) to protect systems against code injection and similar attacks. This technique relies on code interpretation to identify the code section relating to control transfer. Subsequent policing ensures that actual control transfers happening during execution are not in violation of the security policy established earlier. To provide security, program shepherding relies on restricting code origins, restricting control transfers and uncircumventable sandboxing.

Dynamic code-optimization techniques are used to interpret the target binary code. The key underlying principle in this approach is that the code

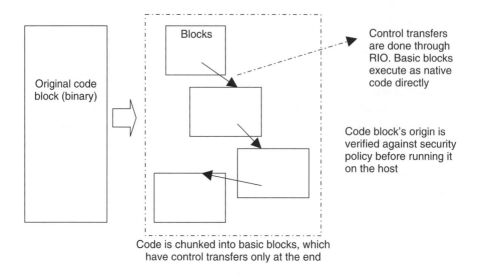

Figure 8.14 Memory firewall.

is separated into blocks (containing a sequence of instructions ending with a single control-transfer instruction). These blocks can be run natively on the host processor. This system prevents injection attacks by controlling all the control transfers in the code (having already identified and partitioned them). Security policies can be defined based on parameters such as code origin, function return, intrasegment jump, intersegment jump, indirect calls, execve, open and so on.

Commercial memory firewall solutions such as SecureCode from Determina are implemented based on the program-shepherding principle.

8.7   Antimalware

Anitmalware, or more popularly antivirus software is pretty much part of the default software stack running on a host. Its main function is to identify and remove any malware that may have compromised a host. Antivirus software today is very versatile and can thwart threats from worms, Trojan horses, spyware and other similar malwares. eTrust, McAfee and TrendMicro are some examples of antivirus solutions.

Antivirus solutions adopt two different approaches, namely signature-based scanning and real-time scanning.

8.7.1 Signature-Based Protection

In this approach, various virus/malware signatures/fingerprints are made available through a dictionary to virus scanners set up at each host [42]. When the antivirus software is able to match a part of a file with the contents of the signature in the dictionary, it tags this file as being infected. An infected file can be cleaned, quarantined or deleted in order to contain the effects of the malware. The effectiveness of this approach largely relies upon the signature dictionary. An up-to-date dictionary can identify more malwares/attacks. But however up-to-date, this approach is inadequate in addressing zero-day attacks (when the virus has not been identified/reported and a signature is not available). Further, there are some viruses which encrypt their own code or use different representations to evade scans by the antivirus scanner.

8.7.2 Real-Time Protection

Recent reports [43] have shown that the traditional signature-based protection is very ineffective when it comes to protecting hosts against unknown viruses. To address these concerns, antivirus systems have included real-time protection, based on system-call interception, to identify attacks from unknown viruses (see Figure 8.15).

User processes don't have the privilege to access system resources directly, and any resource access is mediated by the kernel using system call. By intercepting these system calls, all I/O and resource access can be monitored. Real-time protection [44] uses system-call interception to route the system call through its own filters, which check against policies for any anomalous behavior. The administrator has control over the policies that are used for protection. Real-time protection policies have to be fine-tuned, as they can cause serious performance impact to the host.

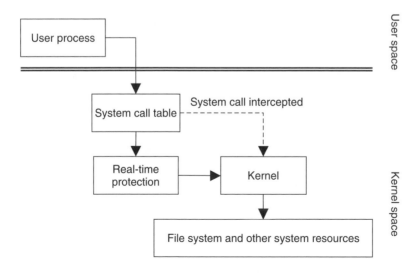

Figure 8.15 Real-time protection.

8.7.3 Heuristics-Based Worm Containment

One approach to providing security against zero-day attack using heuristics. These look at regular patterns of network or file I/O and detect any anomalies [45, 46]. Though this may result in some false positives, as heuristics get better there is a likelihood of better results. Intel vPro uses a similar approach to provide security against worm attacks. In the case of Intel vPro, the solution is tamper-resistant, as the implementation has been done entirely at the hardware level. Figure 8.16 depicts how Intel has implemented this in its platform.

The inline processing unit processes outgoing packets from the host as they pass through the network device driver and network interface. Given the processing time allowed, packets are briefly analyzed and key information packet is put into a cache to be picked up by the sideband processing unit. The sideband processing unit, which is implemented as a hardware component, inspects this packet data in cache and runs it through a heuristics engine to identify any access anomalies. If the heuristic engine identifies any suspicious behavior, it will trigger necessary actions to isolate the node from the network. The administrator can later look at the node and identify whether the host has been compromised. This technique can be used to quickly contain the effects of any malicious code.

8.7.4 Agent Defense

While anitvirus software provides vital security protection to the host either through signature matching or real-time protection, the virus scanners themselves may be disabled when a host is compromised. Unless protection for antivirus is guaranteed, the system may remain vulnerable. One mechanism to guarantee antivirus agent safety is through hardware assists. A firmware-based solution such as Intel's 'system integrity service (SIS)' [47] is OS-independent and can verify

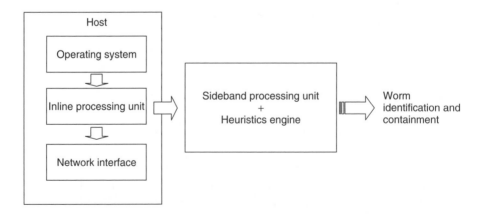

Figure 8.16   Intel's approach to heuristics-based worm containment.

Table 8.1 A summary of host-level solutions.

Attack	Solution	Remarks
Transient code		
Malware	Sandboxing Virtualization Antimalware Memory firewall	Sandboxing and virtualization can play vital roles in containing the effect of malwares at a host level. Antimalware based on signatures is useful in handling known worms and viruses, while systems such as memory firewalls, heuristic-based worm-containment and real-time protection are useful against zero-day attacks. Advances in virtualization, hardware assists for virus protection agents and so on hold lots of promise in providing a more tamper-resistant protection system.
Eavesdropping	Sandboxing Virtualization	The effects of eavesdropping can be contained to a significant extent by using isolation techniques such as sandboxing and virtualization. These techniques are useful for both Web-based software and grid/remote applications.
Job faults	Sandboxing Virtualization	Faults can be effectively contained in isolated environments. These are very useful in grid-like environments.
Resource starvation	Virtualization Resource management	This threat is focused mostly on shared computing facilities such as grid. Virtualization and resource-management solutions can provide effective protection against this threat.
Resident code		
Overflow Privilege escalation Injection	Virtualization Memory firewall Antimalware	Resident-code vulnerabilities are largely due to poor engineering. Viruses, worms and other untrusted/mobile applications running on the host exploit these vulnerabilities. Virtualization, memory firewalls and antimalware can provide reasonable security against these threats. However, all these are inadequate to address them entirely.

the integrity and presence of registered software agents on the host. SIS can detect attacks that tamper with or disable these host-resident software agents. SIS is based on three main concepts, namely agent locality, agent integrity and agent execution state.

Agent locality refers to the memory location where the agent code and data are loaded. Once registered, SIS denies access to any external code which tries to access these confidential bytes. The integrity services module is responsible for ensuring agent locality. SIS verifies agent code for tamper. Any attempt to tamper with the code can be identified by the agent integrity services through integrity manifests. The host agent execution state is communicated to a central integrity measurement manager (IMM) through tamper-proof heartbeats.

8.8 Conclusion

We have looked at some of the host-level security solutions relating to isolation, resource management and host protection. These broadly address the host threats and vulnerabilities we covered in Chapter 4. Table 8.1 summarizes how the solutions described in this chapter address host security and privacy threats.

References

[1] Figueiredo, R., Dinda, P. and Fortes, J. (2003) A Case for Grid Computing on Virtual Machines, Proceedings of ICDCS.
[2] Provos, N. (2003) Improving Host Security with System Call Policies. Proceedings of the 12th USENIX Security Symposium, pp. 257–72, August 2003.
[3] Goldberg, I., Wagner, D., Thomas, R. and Brewer, E. (1996) A Secure Environment for Untrusted Helper Applications. Proceedings of the 6th USENIX Security Symposium, July 1996.
[4] CERT (2002) Vulnerability note VU#176888, Linux Kernel Contains Race Condition via ptrace/procfs/execve. May 2002.
[5] Garfinkel, T., Pfaff, B. and Rosenblum, M. (2004) Ostia: A delegating architecture for secure system call interposition, Proceedings of the Network and Distributed Systems Security Symposium.
[6] Garfinkel, T. (2003) Traps and Pitfalls: Practical Problems in System Call Interposition Based Security Tools. Proceedings of the Network and Distributed Systems Security Symposium, February 2003.
[7] Liang, Z., Venkatakrishnan, V.N. and Sekar, R. (2003) Isolated Program Execution: An Application Transparent Approach for Executing Untrusted Programs, Proceedings of the 19th Annual Computer Security Applications Conference.
[8] Lawton, K. Running Multiple Operating Systems Concurrently on an IA32 PC Using Virtualization Techniques, 2000.
[9] Goldberg, R. (1974) Survey of virtual machine research. *IEEE Computer*, June, 34–45.
[10] Smith, J. (2001) An Overview of Virtual Machine Architectures.
[11] Chen, P. and Noble, B. (2001) When Virtual is Better Than Real, HOT-OS.
[12] Popek, G. and Goldberg, R. (1974) Formal requirements for virtualizable third generation architectures. *Communications of the ACM*, **17**(7), 413–21.
[13] Waldsburger, C. (2002) *Memory Resource Management in VMWare ESX Server*, OSDI.
[14] Whitaker, A., Shaw, M. and Gribble, S.D. (2002) *Denali: Lightweight Virtual Machines for Distributed and Networked Applications*, University of Washington Technical Report 02-02-01.
[15] Barham, P. *et al.* (2003) *Xen and the Art of Virtualization*, SOSP.

[16] Hand, S. *et al.* (2003) *Controlling the XenoServer Open Platform*, OpenARCH.
[17] http://www.parallels.com/r/pdf/wp/pvc/Parallels_Virtuozzo_Containers_WP_an_ introduction_to_os.pdf (Accessed on 12 June 2008).
[18] Tucker, A. and Comay, D. (2004) Solaris Zones, Operating System Support for Server Consolidation, Work in progress reports, USENIX VM.
[19] http://www.parallels.com/en/products/virtuozzo/os/ (Accessed on 12 June 2008).
[20] http://linux-vserver.org/Paper (Accessed on 12 June 2008).
[21] Price, D. and Tucker, A. Solaris Zones: Operating System Support for Consolidating Commercial Workloads, Proceedings of the 18th Large Installation Systems Administration Conference (USENIX LISA '04).
[22] Dike, J. (2000) A User-mode Port of the Linux Kernel, Linux Showcase and Conference 2000 (see also http://user-mode-linux.sourceforge.net/).
[23] Hoxer, H. *et al*, (2002) Implementing a User-mode Linux with Minimal Changes from the Original Kernel 2002, Linux System Technology Conference. Virtual Devices.
[24] Microsoft Corporation (2004) Microsoft Virtual Server 2005 Technical Overview, http://download.microsoft.com/download/5/5/3/55321426-cb43-4672-9123-74ca3af6911d VS2005TechWP.doc. (Accessed on 12-Jun-2008).
[25] Uhlig, R., Neiger, G., Rodgers, D. *et al.* (2005) Intel virtualization technology. *Computer*, **38**(5), 48–56.
[26] AMD Virtualization (2008) http://www.amd.com/us-en/Processors/ProductInformation/0, 30_118_8796_14287,00.html (Accessed on 12-Jun-2008).
[27] Garfinkel, T. and Warfield, A. (2007) *What Virtualization can do for Security*, Vol. **32**(6), login: The USENIX Magazine.
[28] Garfinkel, T. *et al.* (2003) Terra: A Virtual Machine-based Platform for Trusted Computing, SOSP.
[29] Garfinkel, T. and Rosenblum, M. (2003) A Virtual Machine Introspection Based Architecture for Intrusion Detection. Proceedings of the Network and Distributed Systems Security Symposium, February 2003.
[30] Sun Secure Global Desktop (2008) http://www.sun.com/software/products/sgd/index.jsp (Accessed on 12-Jun-2008).
[31] Microsoft Softgrid Application Virtualiztion (2008) http://www.microsoft.com/systemcenter/ softgrid/default.mspx (Accessed on 12-Jun-2008).
[32] VMWare – Virtual Desktop Infrastructure (2008) http://www.vmware.com/products/vdi/ (Accessed on 12-Jun-2008).
[33] Sun Virtual Desktop Infrastructure (2008) http://www.sun.com/software/vdi/index.jsp (Accessed on 12-Jun-2008).
[34] (2007) Performance Impact of Advance Reservations from the Grid on Backfill Algorithms, Sixth International Conference on Grid and Cooperative Computing, 2007. GCC 2007. 16–18 Aug. 2007 pp. 456–61.
[35] Chakrabarti, A., Damodaran, A. and Sengupta, S. (2008) Grid computing security: a taxonomy. *IEEE Security and Privacy*, **6**(1), 44–51.
[36] SystemAdministration Guide: Solaris Containers-Resource Management and Solaris Zones, http://www.sun.com/bigadmin/content/zones/sys-admin-rm.pdf (Accessed on 12 June 2008).
[37] Windows System Resource Manager (2008) http://www.microsoft.com/windowsserver2003/ technologies/management/wsrm/default.mspx (Accessed on 12 June 2008).
[38] Aurema ARMTech Active Resource Management for Application Performance Control (2008) http://www.citrix.com/English/ps2/products/documents_onecat.asp?contentid= 594719&cid=White+Papers#top (Accessed on 12 June 2008).
[39] Entitlement Based Scheduler (2008) http://sourceforge.net/projects/ebs-linux/ (Accessed on 12 June 2008).

[40] Necula, G. (1997) Proof-carrying code. in *24th ACM SIGPLAN-SIGACT Symposium on Principles of Programming Languages*, ACM Press, New York, pp. 106–19.

[41] Kiriansky, V., Bruening, D. and Amarasinghe, S.P. (2002) Secure Execution via Program Shepherding. Proceedings of the 11th USENIX Security Symposium (August 05–09, 2002), pp. 191–206.

[42] The Case for Combined Behavioral and Signature Based Protection (2008) http://www.mcafee.com/us/local_content/white_papers/partners/wp_complete_security_hips.pdf (Accessed on 12 June 2008).

[43] Anti Virus Software Report (2008) http://www.heise.de/ct/08/01/092/ (Accessed on 12 June 2008).

[44] McAfee Entercept (2008) http://www.mcafee.com/us/_tier2/products/_media/mcafee/wp_systemcallinterception.pdf (Accessed on 12 June 2008).

[45] Durham, D., Nagabhushan, G., Sahita, R. and Savagaonkar, U. (2008) *A Tamper-Resistant, Platform-Based, Bilateral Approach to Worm Containment*, Technology@Intel Magazine, http://www.intel.com/technology/magazine/research/worm-containment-1005.htm (Accessed on 12 June 2008).

[46] Savagaonkar, U., Sahita, R., Nagabhushan, G., Rajagopal, P. and Durham, D. (2008) An OS Indepdendent Heuristics Based Worm Containment System, http://www.intel.com/technology/comms/download/worm_containment.pdf (Accessed on 12 June 2008).

[47] Schluessler, T., Khosravi, H., Nagabhushan, G., Sahita, R. and Savagaonkar, U. (2008) Runtime Integrity and Presence Verification for Software Agents, http://www.intel.com/technology/magazine/research/runtime-integrity-1205.htm (Accessed on 12 June 2008).

9

Infrastructure-Level Solutions[1]

9.1 Introduction

In Chapter 5 we discussed the different threats to and vulnerabilities in the enterprise
infrastructure. As we said in that chapter, the IT infrastructure in most enterprises
closely resembles the physical infrastructure – flyovers, buildings, roads, rail and
so on – in the real world. Therefore, it is not surprising that the solutions that have
been proposed to solve infrastructure problems closely resemble those in the real
world as well. For example, similar to passports and visas, certificates, authorization
enforcement systems and firewalls are available in most enterprise IT infrastructures.
In this chapter, we will discuss the solutions that exist for each infrastructure threat
and vulnerability, as well as the potential solutions and research efforts required for
some as yet unresolved problems. We will also provide a recommendation for the
applicability of the solutions at the end of the chapter.

9.2 Network-Level Solutions

If we look at the research in the area of security, that in network security is probably the oldest and most well-established. Solutions such as Secure Socket Layer
(SSL)/Transport Layer Security (TLS), Virtual Private Networks (VPNs), IP Security (IPSec) and so on are well-established and widely deployed. We will briefly
touch upon these in this section. However, several threats like denial-of-service
(DoS), routing attacks and so on are new, and research efforts are still ongoing.
Solutions in the area of wireless networks are evolving, and adoption is taking place.

[1] Contents in this chapter reproduced with permission from Grid Computing Security, Chakrabarti, Anirban, 2007,
XIV, 332 p. 87 illus., Hardcover, ISBN: 978-3-540-44492-3

Distributed Systems Security   A. Belapurkar, A. Chakrabarti, S. Padmanabhuni, H. Ponnapalli, N. Varadarajan
and S. Sundarrajan © 2009 John Wiley & Sons, Ltd

9.2.1   Network Information Security Solutions

Information security covers confidentiality, integrity and authentication, which are generally solved through the combination of encryption and authentication mechanisms. In this subsection, we will briefly cover technologies like SSL/TLS, IPSec and VPN.

9.2.1.1   Secure Socket Layer (SSL)

One of the most popular protocols for securing the transport layer is SSL, whose newer versions are called TLS [1]. SSL/TLS works on top of the Transport Layer Protocol (TCP) and provides security in managing sessions over the transport channel.

SSL Version 2 was deployed with Netscape Navigator 1.1 by Netscape in 1995. Netscape came out with Version 3 a few years later. The Internet Engineering Task Force (IETF) [2] extended the concept to develop TLS.

The protocol works as follows:

(1) The client contacts the server to initiate an SSL/TLS session. In this step the client does not identify itself, but it does mention the set of cryptographic algorithms it can support. In addition, it sends a random-number RC, which will be used to create the session key.
(2) The server replies by sending its certificate to the client. It also sends a random-number RS, which will contribute toward the creation of the session key.
(3) The client then verifies the certificate, extracts the public key of the server and selects a random number S. In addition, the client also computes K, which is the master secret computed as a function of RC, RS and S.
(4) The client sends S and the hash of K encrypted with the server's public key.
(5) Subsequently, all the data sent over the SSL/TLS channel is encrypted with the session key K.

It is to be noted that the SSL/TLS protocol defined above helps the client to authenticate the server. However, the server cannot authenticate the client. As a protocol, SSL/TLS allows the option for mutual authentication, where the server can authenticate the client if the client possesses the required certificate. However, in most cases, if such an authentication is required the client sends its user name and password encrypted with the server's public key.

9.2.1.2   IP Security (IPSec)

IPSec [3, 4] is a method proposed to solve attacks through interaction with the network layer. The principal feature of IPSec, which enables it to support a variety of application scenarios, is that it can encrypt or authenticate all traffic at the

Internet protocol (IP) level. Thus, all distributed applications, including remote login, client/server, e-mail, file transfer, Web access and so on, can be secured.

An organization maintains local area networks (LANs) at dispersed locations. Traffic on each LAN does not need any special protection, but the devices on a LAN can be protected from an untrusted network with firewalls. Since we live in a distributed and mobile world, the people who need to access the services on each of the LANs may be at sites across the Internet. These people can use IPSec protocols to protect their access. IPSec protocols can operate in networking devices, such as a router or firewall that connects each LAN to the outside world, or else directly on the workstation or server. The user workstation can establish an IPSec tunnel with the network devices to protect all subsequent sessions. After this tunnel is established, the workstation may have many different sessions with the devices behind the IPSec gateways. The packets going across the Internet will be protected by IPSec but will be delivered on to each LAN as a normal IP packet.

IPSec has the following main components:

(1) *Two security mechanisms:* an authentication-only function, referred to as Authentication Header (AH) [4], and a combined authentication and encryption function called Encapsulating Security Payload (ESP) [5]. These provide the basic security mechanisms within the IP.
(2) *Security Associations (SAs):* these represent an agreement between two peers on a set of security services to be applied to the IP traffic stream between them.
(3) *Key management infrastructure:* sets up SA between two communicating peers.

Both AH and ESP security mechanisms involve adding a new header to the IP packet, and the header is added between the original IP header and the layer 4 (network layer) header. In this way, only the two IPSec peers will have to deal with the additional headers, thus letting legacy routers handle IPSec packets just like normal IP packets. This feature lets far fewer IPSec-compliant devices on the Internet, making its deployment easier. IP AH and IP ESP may be applied alone or in combination. Each function can operate in one of two modes: transport mode or tunnel mode. With transport mode, AH or ESP is applied only to the packet payload, while the original IP packet header remains untouched. The AH or ESP header is inserted between the IP header and the layer 4 header, if there is any. In tunnel mode, AH or ESP is applied to the entire original IP packet, which is then encapsulated into a new IP packet with a different header.

For VPN, both authentication and encryption are generally desired, because it is important both to (i) assure that unauthorized users do not penetrate the VPN and (ii) assure that eavesdroppers on the Internet cannot read messages sent over the VPN. Because both features are generally desirable, most implementations are likely to use ESP rather than AH. However, by providing both AH and ESP, IPSec provides implementers with flexibility in terms of performance and security. This flexibility is also extended to the key-exchange function, where both manual and automated key-exchange schemes are supported.

9.2.1.3 Virtual Private Networks

To carry out business-related activities, it has become absolutely essential for employees, contractors and business managers to be able to access confidential resources and communicate them across geography. It is quite common for business executives to log on and access resources using laptops while traveling, or when they are in client or business locations. Since the communications generally take place over a public network, confidentiality, authentication and integrity are very important. One prominent communication service which provides these and allows access to resources anywhere and anytime is VPN. Before the advent and popularity of VPN technologies, private networks were created using permanent links between corporate sites. VPN technologies extend this concept by providing virtual networks that are dynamic and setting up connections according to organizational needs. Unlike traditional corporate networks, VPNs do not maintain permanent links between end-points. Rather, the connection is torn down as soon as it is not required, resulting in bandwidth savings. VPN technologies are cost-effective alternatives to completely private networks, allowing different parties to come together and share resources in a secure manner.

There are mainly two different types of VPN technology. These are Layer 2 VPN Service (L2VPN) and Layer 3 VPN Service (L3VPN).

Layer 2 VPN service (L2VPN)
In L2VPNs, the provider extends layer 2 services to the customer sites. A key property of L2VPNs is that the provider is unaware of layer 3-specific (network layer) VPN information. The customer and the provider do not exchange any routing information with each other. Forwarding decisions in the provider network are based solely on layer 2 (data-link layer) information such as message authentication code (MAC) address, ATM VC identifier, multiprotocol label switching (MPLS) label and port number. Currently, two different approaches to L2VPNs are described in the literature; Virtual Private Wire Service (VPWS) and Virtual Private LAN Service (VPLS) [6]. The major difference between the two is that the VPWS provides VPN service between two sites, while VPLS provides a service across multiple sites. The VPWS approach can be regarded as a generalized version of the traditional leased-line service, in which the sites are connected in a partial or full mesh. The VPLS approach emulates a LAN environment, where a site automatically gains connectivity to all the other sites attached to the same emulated LAN.

Layer 3 VPN service (L3VPN)
In L3VPNs, the provider offers layer 3 (network layer) connectivity, typically IP, between the different customer sites. At present, there are two dominating L3VPN

approaches, (Border Gateway Protocol) BGP/MPLS VPN [7] and Virtual Router (VR) [8]. Both approaches concentrate the VPN functionality at the edge of the provider network (Provider Edge (PE) nodes) and hide VPN-specific information from the provider core nodes, to improve scalability. In the BGP/MPLS VPN approach, a routing context is represented as a separate routing and forwarding table in the PE. Each PE node runs a single instance of a BGP variant called Multiprotocol BGP (MPBGP) [9] for VPN route distribution across the core network. PE nodes use MPLS labels to keep VPN traffic isolated and transmit packets across the core network in tunnels. The tunnels are not necessarily MPLS tunnels; they can be of any type, such as IPSec. If a tunnel type other than MPLS is used, the only nodes that need to know about MPLS are the PEs. Any routing protocol can run between the Customer Edge (CE) nodes and the PEs, but in practice the customer must use the routing protocol chosen by the provider. In the VR approach, PE nodes have one VR instance running for each VPN context. A VR emulates a physical router and functions exactly like one. VRs belonging to the same VPN are connected to each other via tunnels across the core network.

9.2.2 Denial-of-Service Solutions

Solutions proposed in the literature for DoS attacks can be broadly categorized as (i) preventive and (ii) reactive. Preventive DoS solutions take precautionary steps in preventing DoS attacks. A wide range of solutions have been proposed, but this problem remains an open one. The reactive solutions aim at identifying the source of attacks. This is very important because attackers spoof their addresses, thus techniques are needed to trace them.

9.2.2.1 Preventive DoS Countermeasures

The preventive DoS techniques are used to detect and reduce the effectiveness of attacks. In this section we will talk about some of the methods that have been proposed to detect and prevent DoS attacks, namely filtering, location-hiding and throttling techniques.

Packet filtering

All preventive DoS-detection techniques are based on some prior information, on the basis of which the filtering is carried out. A few filtering techniques are described in Cisco White Papers [10]. One of the most common methods for detecting and preventing potential attacks is to use egress filtering. Egress filtering refers to the practice of scanning the packet headers of IP packets leaving a network (egress packets) and checking to see if they meet certain criteria. If the packets pass the criteria, they are routed outside of the subnetwork from which they originated. If they do not pass, the packets will not be sent to the intended target. Since one of the features of distributed denial-of-service (DDoS) attacks is spoofed IP addresses, there is a good probability that the spoofed source address of DDoS attack packets will not represent a valid source address of the specific subnetwork. If the network administrator places a firewall or packet sniffer in the subnetwork to filter out any traffic without an originating IP address from this subnet, many DDoS packets with spoofed IP source addresses will be discarded, and hence neutralized. This is a common technique and has been deployed as a defense mechanism in routers [10]. It is to be noted that these types of measure can minimize attacks to some extent, but can in no way guarantee an absolute defense against them.

Similar to egress filtering, different ingress filtering mechanisms have also been proposed and implemented. In these types of mechanism, the filtering is done on all packets coming into the network. In [11], the authors have presented techniques for preventing DoS attacks through filtering. One such technique is called distributed packet filtering (DPF), where the decision to drop or accept the packet is made based on the incoming packet interface. The route information plays a major role in determining whether a packet should be dropped or not. The route information stored at each node indicates the source address and the corresponding interface that the packet is supposed to come from.

Application filtering
Just filtering packets does not provide enough protection from XML denial-of-service (XDoS) attacks, for which it is necessary to be able to understand Extensible Markup Language (XML) documents. XML-level firewalls examine the received Simple Object Access Protocol (SOAP) messages or native XML messages and, once the target Web service is resolved, apply a stored security policy based on the target address, originating caller identity, message content and, in some cases, the successful execution of prior policies. Several companies, like Reactivity, have developed such firewalls. Most of the common XDoS attacks, such as entity-expansion attacks, can be filtered by adding specific policies at the XML firewall level. It is to be noted that this type of filtering has a significant effect on performance as complex policies need to be applied to the incoming XML messages. Also, newer attacks are constantly being invented in the growing field of Web services and the filtering technique needs to keep up with these.

Location hiding
In this type of prevention mechanism, the actual location is hidden from the end users, preventing attack from taking place. An architecture built on this principle is called Secure Overlay Service (SOS) [12]. The goal of the architecture is to allow communication between a confirmed user and a target. The target selects a subset of nodes N that participate in the SOS overlay to act as *forwarding proxies*. The filter *only allows* packets whose source address matches the address of some overlay node *n* in N. It is assumed that the set of nodes that participate

in the overlay is known to the public, and to the attacker as well. Attackers in the network are interested in preventing traffic from reaching the target. By hiding the actual target, SOS reduces the effectiveness of DoS attacks. However, there are some concerns: first, the SOS architecture may cause additional latency because of the large amount of forwarding and routing that takes place. In some applications, this may be really critical. Second, the architecture assumes that the attack is coming from outside and does not concentrate on insider attack.

Throttling

One proposed method to prevent servers from going down is to use max−min fair server-centric router throttles [13]. This method sets up routers that access a server with logic to adjust (throttle) incoming traffic to levels that it will be safe for the server to process. This will prevent flood damage to servers. Additionally, this method can be extended to throttle DDoS attacking traffic versus legitimate user traffic for better results. This method is still in the experimental stage. However, similar techniques to throttling are being implemented by network operators. The difficulty with implementing throttling is that it is still hard to decipher legitimate traffic from malicious traffic. In the process of throttling, legitimate traffic may sometimes be dropped or delayed and malicious traffic may be allowed to pass to the servers. One of the projects which use throttling as a means of mitigating DoS attacks is the D-WARD project from UCLA. D-WARD is a DDoS defense system deployed at source-end networks, which autonomously detects and defeats attacks originating from these networks. It includes observation and throttling components, which can be part of the source router, or a separate unit which interacts with the source router to obtain traffic statistics and install rate-limiting rules. The observation component monitors two-way traffic at a flow granularity to detect attack. Flow classification, connection classification, the TCP normal traffic model, the ICMP normal traffic model and the UDP normal traffic model are used to differentiate the malicious flow and the legitimate flow. Once the attack flow is found, the misbehavior flow comes under the control of rate-limiting rules. D-WARD can detect some attacks at the source-edge network, and it attempts to determine outgoing attack traffic. But since there is no coordination among instances of agents, the detection may be error-prone. Source-end defense is a promising scheme that can be applied in the active defense system. However, it faces lots of challenges, such as detection sensitivity, agent coordination and liability. When the defense system is deployed in the source end there are fewer strong signals to indicate the attack than at the victim end, at which there are usually apparent signals, such as high volume of network traffic. So a high sensitivity is essential for source-end defending.

Intrusion detection systems (IDS)

Intrusion Detection System (IDSs) as just a preventive DoS measure. IDSs [14, 15] consist of detectors that detect attacks based on a set of policies and information. In principle, they work similarly to alarm systems implemented in buildings and apartments for protection against burglars. In [14], the authors have grouped IDSs into two main categories: *anomaly detection systems* and *signature detection systems*. In the former, an intrusion is detected based on abnormalities of system behavior. The detector forms an opinion based on the normal behavior of the system, determined by long-term observation and system policies. In the latter, an intrusion is detected based on a specific signature or a model. It is to be noted that the signature is based on long-term information about the intrusion behavior.

Several grid-based IDSs have been conceived, designed and implemented. Most consist of a set of sensors which are able to monitor the state of the grid systems. The information supplied by the sensors is collected and analyzed by IDSs like SNORT [16]. It is then logged through an interface to query it, and suitable alarms and action mechanisms are provided. Grid-based systems described in the literature include SANTA-G [17], which uses SNORT as its IDS and R-GMA for querying the monitored information, and GIDA [18], which uses a similar structure. IDS on the Oracle 10G database is provided in [19]. Integrated Access Control and Intrusion Detection (IACID) [20] from USC provides a grid-based IDS with separate network and host IDSs.

9.2.2.2   Reactive DoS Countermeasures

Reactive techniques aim at identifying the attacker after the attack has been completed. This is an active area of research because the current identification techniques are totally manual and may span months. The current solutions can be broadly categorized as link testing, logging, ICMP traceback and IP traceback.

Link testing

This technique involves iteratively checking the upstream link until the source is reached. This type of identification technique assumes that the attack remains active after the completion of the trace. One type of link-testing approach is called input debugging, where routers develop an attack signature based on some attack pattern. The victim informs the operator about the signature and the operator then checks the packets and iteratively carries out this process. This is employed in some routers now, though the process is time-consuming. Another suggested link-testing technique is controlled flooding [21], in which the victim floods all the links, based on the assumption that the packet drop taking place from an attacked link is much more than from any other link. This technique suffers from being a mode of DoS attack itself.

Logging

A simple technique has been suggested in [22], where logging of data packets is done at key routers. Traceback is carried out by using data-mining techniques. Another interesting work in this area is reported in [23], where the authors have presented a hash-based technique for IP traceback that generates audit trails for traffic within the network. The origin of packets can be traced back to the source based on these audit trails. In Source Path Isolation Engine (SPIE) architecture (Refer to the below Figure), each router consists of a data-generation agent (DGA) which computes the hash of the packets and stores them in a Bloom filter. The information is flushed after every time interval t, where t is a design parameter. As soon as the attack is detected, the SPIE Traceback Manager (STM) calculates the attack signature of the packet or packets used for the attack. It then contacts the centralized SPIE Collection and Reduction Agent (SCAR). SCAR polls DGAs for the information stored, creates a local attack graph, and sends the information back to the STM. The STM then assembles the local graphs, plugs holes in the graphs and finally creates the traceback information. This technique suffers from scalability problems, as enormous resources are required to carry out logging-based identification. Another negative, which is associated with all traceback schemes, is that it can only trace back to a single attacker. Since most attacks are carried out using reflectors, traceback schemes can rarely be used.

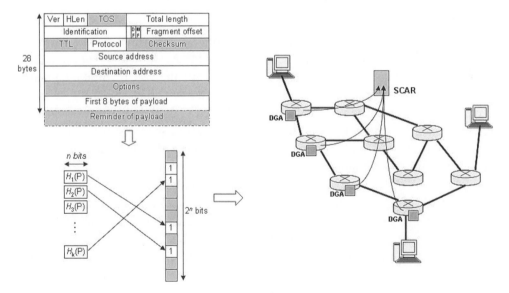

ICMP traceback

In the Internet draft [24], the author has proposed a scalable technique where each router stores packets with a low probability. Whenever a packet is stored, the router sends an ICMP traceback message toward the destination. When attacked, the destination can trace back to the source based on the router ICMP messages. This scheme has a problem as the ICMP messages can themselves be used to cause DoS attacks.

IP traceback

One of the earliest efforts to identify the source of a packet through IP traceback was made in [25] through probabilistic marking of packets at each router. In this technique, a router marks any packet flowing through it with a very small probability. With a sufficient number of packets (in the case of DoS attacks), the destination can retrace the attack path. The scheme introduces a huge amount of overhead to the packets, but this can be reduced when only one field is reserved for marking and the information gets overwritten. The attack path, in this case, is retraced by a number of packets marked at each router. This type of traceback is called node sampling. The authors of [25] also introduced a concept called edge sampling, where in addition to the information of the node, the distance to the node is also maintained. The schemes were further extended in [26], where the authors showed that using partial network information, the number of packets required to trace back could be substantially reduced.

9.2.3   DNS Solution – DNSSEC

DNSSEC provides authentication and integrity to the Domain Name System (DNS) updates. All of the DNS attacks (mentioned in Chapter 5) are mitigated with the addition of data-origin authentication, and transaction and request authentications. The authentications are provided through the use of digital-signature technology. The digital signature contains the encrypted hash of the resource record set (RR set). The recipient can then check the digital signature against the received data. To make the DNSSEC proposals valid, secure servers and secure client environments must be created. Moreover, DNSSEC is unable to provide security against information leakage as it is mainly concerned with authentication. The primary goal of DNSSEC is to provide authentication and integrity for data received from the DNS database. This is done via digital-signature schemes based on public-key cryptography. A possible approach is to sign each DNS message. The general idea is that each node in the DNS tree is associated with a public key of some sort. Each message from DNS servers is signed under the corresponding private key. It is assumed that one or more authenticated DNS root public key is publicly known. These keys are used to generate certificates on behalf of the top-level domains; that is, these keys are used to generate a signature that binds the identity information of each top-level domain to the corresponding public key. The top-level domains sign the keys of their subdomains, in a process where each parent signs the public keys of all its children in the DNS tree.

9.2.4   Routing Attack Solutions

Another important attack possible against network infrastructure is the routing attack. As mentioned in Chapter 5, routing attacks are mostly concerned with the routing packets flowing between the routers. Attacks can be of two types: link attacks and router attacks. The solutions for the two are described below.

9.2.4.1   Link-Attack Solutions

Most routing protocols employ robust updates between neighbors [27, 28] by using acknowledgments. Link attacks are detected in those cases. However, if links interrupt selectively, it is possible to have unsynchronized routing tables throughout the network. The after-effects of such routing tables are looping and DoS. Unsynchronized routing tables can also be created if a router drops the updates but sends an acknowledgment. The problem of routers dropping routing updates selectively has not been studied in the literature.

Digital signatures [29] are used to provide integrity and authenticity to messages. With digital signatures, the sender signs packets with its private key, and all nodes can verify the signature based on the sender's public key. In this case, the routing updates increase by the size of the signature (typically between 128 and 1024 bits). This is a viable solution in link-state routing protocols, since the LSAs are transmitted infrequently. It is also proposed as a solution for distance-vector protocols. Distance-vector protocols suffer from excessive bandwidth consumption as the distance vectors are exchanged quite frequently. Therefore, the addition of extra overhead in the form of a digital signature has been looked upon by the research community with concern. Efforts have been undertaken to reduce the overhead through the use of efficient digital signatures [29]. Another problem with this approach is that it relies on the existence a public-key infrastructure (PKI) for its functioning. In absence of a PKI, the proposed solutions are not viable.

Sequence information is used to prevent this attack [30]. Sequence information can take the form of sequence numbers or time stamps. An update is accepted as valid if the sequence number in the packet is greater than or equal to the sequence number of the previously-received update from the same router. This solves the problem of replication, but the packets within the same clock period can be replayed if the time stamp is used as sequence information. No remedy has been found for this problem. However, this problem has limited effect as it can be employed only if a router sends multiple updates within the same time period.

9.2.4.2   Router-Attack Solutions

The solutions proposed for router attacks in link-state protocols can be categorized into two types: intrusion detection and protocol-driven. Use of intrusion-detection techniques has been suggested as a mechanism to detect router attacks [31]. In these techniques, a centralized attack-analyzer module detects attacks based on possible alarm-events sequences. Using an attack-analyzer module in an Internet scenario is not a scalable solution. In a protocol-driven solution, the detection capability is embedded in the link-state protocol itself. In [32], Secure Link State Protocol (SLIP) is proposed where attack detection capability is incorporated in the routing protocol itself. A router does not believe an update unless it receives 'confirmation' link-state update from the node supporting the questionable link. However, the solution is not complete as it works only in symmetric networks where both nodes supporting the link can identify the change in the link state. It also makes an assumption that no malicious collusion exists in the network.

In [33] the authors have proposed a validation scheme (called consistency-check CC algorithm) through the addition of predecessor information in the distance-vector update. Whenever a node receives the distance vector from its neighbors, it carries out CC by tracing the path from each destination. Any inconsistency in the update arising out of a router attack can be identified by the CC alogrithm consistent. However, the CC algorithm is unable to detect router attacks when a malicious router changes the update intelligently, keeping the network topology in mind. Though this is an important issue, not much work is being done to solve the various problems associated with it, and hence it requires significant research attention.

As mentioned in Chapter 5, wireless technologies are gaining in prominence among both individuals and enterprises. However, there are security issues and vulnerabilities which need to be tackled. In this section we provide an overview of the wireless network security solutions. One of the most popular wireless standards is 802.11, which was standardized way back in 1997.

9.2.4.3 Wireless Equivalent Privacy (WEP)

One of the earliest and most popular security mechanisms in wireless LANs was Wireless Equivalent Privacy (WEP). WEP has three goals to achieve for wireless LAN: confidentiality, availability and integrity [34]. It uses encryption to provide confidentiality. The encryption process is only between the client and the Authentication Protocol (AP), meaning that packet transfers after the AP (wired LAN) are unencrypted. WEP uses RC4 for encryption purposes. Since RC4 is a stream cipher, it needs a seed value to start its key-stream generator. This seed is called the initialization vector IV. The IV and the shared WEP key are used to encrypt/decrypt transferred packets. In the encryption process, the integrity-check (IC) value is computed and attached to the payload, then the payload is XORed, with the encryption key consisting of two sections (IV and WEP Key). The packet is then forwarded, with the IV value sent in plaintext. WEP uses CRC (cyclical redundancy checking) to verify message integrity. On the other side (receiver−AP) the decryption process is the same but reversed. The AP uses the IV value sent in plaintext to decrypt the message, by joining it with the shared WEP key.

One of the major reasons for WEP weaknesses is its key length. WEP has a 40-bit key, which can be broken in less than five hours using parallel attacks with the help of normal computer machines. This issue urged vendors to update WEP from using a 40-bit to a 104-bit key; the new release is called WEP2. This update helped to resolve some security issues with WEP.

The main disadvantage of WEP, however, is the lack of key management. Some SOHO (small office/home office) users never change their WEP key, and once it is known the whole system is in jeopardy. In addition to that, WEP does not support mutual authentication. It only authenticates the client, making it open to rogue AP attacks. Another issue is the use of CRC to ensure integrity. While CRC is a good integrity-provision standard, it lacks a strong cryptography feature. CRC is known to be linear. By using a form of induction, knowing enough data (encrypted packets) and acquiring specific plaintext, the WEP key can be resolved.

802.11x: EAP over LAN

The 802.1x standard was designed for port base authentication for 802 networks. 802.1x is not concerned with the type of encryption technique employed, it is only used to authenticate users. Extensible Authentication Protocol (EAP) was designed to support multiple authentication methods over point-to-point connections without requiring IP (RFC 3748). EAP allows any of the encryption schemes to be implemented on top of it, adding flexibility to the security-design module. EAP over LAN (EAPOL) is EAP's implementation for LANs. The 802.1x framework defines three ports or entities: Supplicant (the client that wants to be authenticated), Authenticator (the AP that connect the supplicant to the wired network) and Authentication Server (AS) (which performs the authentication process from the supplicant, based on their credentials). The AS in the 802.1x framework uses Remote Authentication Dial-In User Service (RADIUS) protocol to provide authentication, authorization and accounting (AAA) service for network clients. The protocol creates an encrypted tunnel between the AS and the Authenticator (AP). Authentication messages are exchanged inside the tunnel to determine whether the client has access to the network.

802.11i standard

The 802.11i (released June 2004) security standard is the latest solution to plug the security holes in wireless networks through authentication, integrity and data transfer. 802.11i supports two methods of authentication. The first is the one described before, using 802.1x and EAP to authenticate users. For users who cannot or do not want to implement this, the second method was proposed, using per-session key per device. This method is implemented by having a shared key (like the one in WEP), called the Group Master Key (GMK). The GMK is used to derive the Pair Transient Key (PTK) and the Pair Session Key (PSK), which do the authentication and data encryption. To solve the integrity problem with WEP, a new algorithm named Michael is used to calculate an 8-byte IC called message integrity code (MIC). Michael differs from the old CRC method by protecting both the data and the header. It implements a frame counter, which helps to protect against replay attacks [35]. To improve data transfer, 802.11i specifies three protocols: Temporal Key Integrity Management (TKIP), Counter with Cipher Block Chaining Message Authentication Code Protocol CCMP and Wireless Robust Authenticated Protocol (WRAP). TKIP was introduced as an ad hoc solution to WEP problems. One of the major advantages of implementing TKIP is that one does not need to update the hardware to run it; simple firmware/software upgrade is enough. Unlike WEP, TKIP provides per-packet key mixing, a message IC and a rekeying mechanism. TKIP ensures that every data packet is sent with its own unique encryption key. TKIP is included in 802.11i mainly for backward compatibility. WRAP is the LAN implementation of the popular Advanced Encryption Standard (AES) [36]. It was ported to wireless to get the benefits of AES encryption. WRAP has intellectual property issues, where three parties have filed for its patent. This problem caused IEEE to replace it with CCMP. CCMP is considered the optimal solution for secure data transfer under 802.11i. CCMP uses AES for encryption. The use of AES will require a hardware upgrade to support the new encryption algorithm.

9.2.5   Comments on Network Solutions

The network is one of the most critical components of any enterprise infrastructure. As mentioned in Chapter 5, there are four main types of attack: DoS attack, DNS attack, routing attack and wireless attack. Some generic solutions like SSL/TLS, VPN and IPSec are used for information security, in conjunction with many others. DoS attacks still require a lot of research before a holistic solution will become available. However, preventive solutions like application and egress filtering are of benefit in most cases. The proposed DNSSEC solution can prevent most DNS attacks. Routing attacks, especially consistency-based attacks, are difficult to detect, and not many solutions have been implemented. Wireless attacks can also be dangerous in enterprise scenarios, as more and more wireless applications are being deployed. Several solutions, like WEP, 802.11x and 802.11i, have been proposed. In Table 9.1 we provide a snapshot of different network-level solutions and their effectiveness.

9.3   Grid-Level Solutions

As mentioned in Chapter 5, grid security issues can be categorized into architecture issues, infrastructure issues and management issues. In this section, we will briefly talk about the different solutions pertaining to each issue. Please refer to [32] for details about each solution.

Table 9.1 Overview of network solutions.

Category	Solution	Problems solved	Comments
Information security solutions	SSL	Confidentiality, authentication	Very popular for application-level security
	VPN	Confidentiality, authentication	Very popular for remote access
	IPSec	Confidentiality	Important for network security
DoS solutions	Preventive	Preventing DoS attack	Filtering and IDS solutions have been implemented and have proved to be a deterrent in some cases
	Reactive	Identifying the attacker	More research is needed as the solutions, like packet marking and so on, are mostly research prototypes
DNS solution	DNSSEC	Authentication and confidentiality of DNS data	Most of the important DNS attacks like cache poisoning can be limited
Routing attack solutions	Link attack	Confidentiality, fabrication, modification of routing message exchange	Most of the link-level attacks are mitigated using standard cryptographic techniques
	Router attack	Router consistency attacks	More research is needed in this area as the solutions are mostly research prototypes
Wireless attacks	WEP	Confidentiality	Popular, but security vulnerabilities and not robust enough
	802.11x	Authentication	Multiple authentication over point-to-point connections
	802.11i	Authentication, integrity and confidentiality	Robust solution

9.3.1   Architecture Security Solutions

When we look at architecture, issues regarding information security, authorization and service security need to be looked at. In this subsection, we will look at each in turn. We will talk about Grid Security Infrastructure (GSI), which is a proposed standard in the area of grid information security. We will also describe two different types of grid authorization system, namely virtual organization (VO)-level authorization and resource-level authorization.

9.3.1.1   Information Security Solution – Grid Security Infrastructure (GSI)

Grid computing provides a virtualized view of the underlying grid resources. Such a virtualization encompasses the security requirements. Therefore, there is a need for virtualization of security semantics to use standardized ways of segmenting security components like authentication, access control, confidentiality and so on, and to provide a standardized way of enabling the federation of multiple security mechanisms. This requires a loosely-coupled platform-independent model of securing applications within and across organizations. The question arises, which paradigm should be involved in implementing such an architecture?

Web services has the ability to deliver integrated, interoperable services. Since Web services is gradually becoming a default and an industry standard, the OGSA grid computing model uses Web services as a model reference. Confidentiality, integrity, policy management and trust management are also integral to Web services, so GSI integrates the Web services standards like WS-Security, WS-Policy, WS-Trust and so on in the specification. However, GSI does not exclude TLS like SSL on top of HTTP or HTTPs. Users are free to use HTTPs, which provides confidentiality, integrity and authentication. If there is a need to traverse multiple intermediaries, WS-Security can be used in conjunction with XML encryption, signatures and so on.

- *Authentication:* the most prevalent mechanism of authentication in a GSI-based grid is the certificate-based authentication mechanism, where a public-key infrastructure (PKI) is assumed which allows the trusted authority to sign information to be used for authentication purposes. In addition to the certificate-based mechanism, Kerberos and password-based mechanisms have also been implemented.
- *Delegation in GSI:* another very important requirement for a grid-based security system is delegation, where another entity gets the right to perform some action on the user's behalf. This is especially important in grid because of the possibility of multiple resources being involved in grid-based transactions. It may be unnecessary or very expensive to authenticate each and every time a resource is accessed. If the user issues a certificate allowing a resource to act on their behalf then the process will become a lot simpler. Such a certificate is called

a proxy certificate. A proxy is made up of a new certificate containing two parts: a new public and a new private key. The proxy certificate has its owner's identity, with a slight change to show that it is a proxy. The certificate owner, not a CA, will sign the proxy certificate. It will have an entry with a timestamp, which indicates at what time it expires – by default it has a short-term validity period of, say, a few hours.

9.3.1.2 Grid Authorization Systems

Authorization systems can be divided into two main categories: VO-level systems and resource-level systems. VO-level systems have a centralized authorization system which provides credentials for users to access resources. Resource-level authorization systems, on the other hand, allow access to resources based on the credentials presented by the users.

VO-level systems
VO-level grid authorization systems provide centralized authorization for an entire VO. This type of system is necessitated by the presence of a VO which has a set of users and several resource providers (RPs). Whenever a user wants to access certain resources owned by an RP, they obtain a credential from the authorization system, which allows them certain rights. They present the credentials to the resource in order to gain access to it. In this type of system, the resources hold the final right to allow or deny access to a user. The Community Authorization Service (CAS) [37], developed as part of the Globus toolkit, is an example of a VO-level authorization system. CAS (see Figure 9.1) looks at the problem

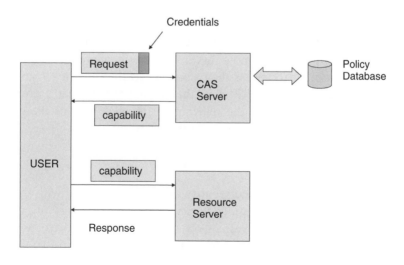

Figure 9.1   High-level working of CAS.

of scalable representation and enforcement of access policies within distributed virtual communities. Such communities may comprise many subcommunities, each participating as a resource provider and/or resource consumer. The problem of authorization is handled using a trusted third party called the CAS server, which is responsible for managing the policies and governing access to the community's resources. Other examples are the Virtual Organization Membership Service (VOMS) [38] and the Enterprise Authorization and Licensing System (EALS) [39].

Resource-level authorization system

Unlike the VO-level authorization systems, which provide a consolidated authorization service for the VO, the resource-level authorization systems implement the decision to authorize access to a set of resources. Therefore, VO-level and resource-level authorization systems look at two different aspects of grid authorization. As will be shown later in this chapter, the two authorization systems complement each other, and can be implemented together to provide a holistic authorization solution. Examples of resource-level authorization systems are Akenti [40], Privilege and Role Management Infrastructure Standards Validation (PERMIS) [41] and the GridMap system.

Akenti was developed by Lawrence Berkeley National Laboratory (LBNL). Though developed with Web resources in mind, the concept was later extended to include resources in a grid-computing VO set-up. The Akenti model consists of resources, including Web resources and distributed-grid resources, that are accessed via a resource gateway (or a Policy Enforcement Point (PEP)) by a set of users who are part of the VO. The model also assumes that each resource may have multiple stakeholders, each with a set of access constraints on the resource set. The Akenti system allows users to access the resource(s) based on their identity and the access policy set on the resources by the resource stakeholders. The stakeholders express their access constraints through a set of self-signed certificates which are known to be stored in a secure remote server. The certificates express the attributes a user must have in order to access the resources. At the time of resource access, the resource gatekeeper or the PEP asks the Akenti server what access the user has. The Akenti server finds all the relevant certificates, verifies the certificates and returns the decision to grant access to the user.

9.3.1.3 Grid Service Security

As mentioned in Chapter 5, service security in grid is concerned with DoS attacks and quality-of-service (QoS) violation attacks. While DoS attack solutions are similar to those described in Section 9.2.2, QoS violation solutions detect and mitigate service-level-agreement (SLA) violations. The solutions can be of two main types: monitoring and auditing systems, and protocol-specific solutions. As part of the monitoring and auditing discussions, we will mention the network-monitoring mechanisms and grid-auditing systems like GridBank. We will also briefly touch upon an interesting work in detecting packet dropping called WATCHERS, based in the University of California (UC) at Davis.

SLA-violation detection in networking infrastructure

In this type of mechanism, SLA violations in networks are detected by monitoring packets and measuring the delays and packet losses they suffer [42–45]. Delay-bound guarantees made by a provider network to user traffic flows are for the delays experienced by the flows between the ingress and egress routers of the provider domain. Delay measurements use delay of either real user traffic or injected traffic. The first approach is intrusive because encoding timestamps into the data packets requires changing the packets at the ingress and rewriting the original content at the egress, after appropriate measurements. The second approach is nonintrusive in that one can inject probe packets with the desired control information, and an egress router can recognize such probes, perform measurements and delete the probes from the traffic stream. Packet-loss guarantees made by a provider network to a user are for the packet losses experienced by its conforming traffic inside the provider domain. To compute the loss ratio, the number of packet drops, as well as the number of packets traversing the domain, is required. Loss ratio is defined as the ratio of the number of packet drops within the domain to the total number of packets passing through the domain. Core routers can detect the number of packets dropped, and edge routers can compute the number of packets traversing the domain. This loss-measurement mechanism can be called the core-assisted scheme for loss measurement. An alternative mechanism uses stripe-based probing to infer loss characteristics inside a domain. In the stripe-based mechanism, a series of packets or 'stripes' are sent which do not introduce intermediate delays.

WATCHERS project

WATCHERS [46], from UC Davis, was proposed to detect and react to routers that maliciously drop or misroute packets. WATCHERS is based on the principle of packet-flow conservation; that is, the number of incoming packets for a router, excluding those destined to it, should be the same as the number of outgoing packets, excluding those generated by it. In order to validate the conservation law, multiple decentralized counters are periodically and synchronously exchanged among neighbors of the target suspected router. Subsequently, each neighboring router runs a validation algorithm to diagnose the health condition of the target router. Furthermore, WATCHERS is robust against Byzantine faults. While WATCHERS theoretically offers an interesting way to deal with malicious packet dropping, it cannot handle the packet dropping problems in today's Internet effectively. First, the number of messages for counter value exchanges can be very large. Second, the principle of packet-flow conservation does not hold 'deterministically' for today's Internet environment. For instance, an innocent router might drop packets for good reasons, such as preventive congestion control or having insufficient resources to keep all incoming packets. Though WATCHERS may not be a viable solution in an Internet environment, it could be a useful tool in a controlled grid system, where the number of packets dropped is not as large as it is in the Internet.

Grid accounting systems

Researchers in the grid community have started to realize the importance of QoS in grid computing systems. Several research projects, such as GridBank [47] from the University of Melbourne and SweGrid [48] from the Royal Institute of Technology in Sweden, have tried to address this issue through their accounting and auditing systems. The former is more of an accounting system, with charging and payment modules. It also includes a service cost negotiation (e.g. $ per hour) which is carried out by the Grid Resource Broker (GRB). GridBank issues Grid-Cheques (similar to credentials) for the service consumers, and Grid Resource Meters gather resource-specific usage information, to be used for charging purposes. The SweGrid system goes beyond being just an accounting system, with SLA negotiation, monitoring and management. SLAs are negotiated through the negotiation phase and monitored using agents. Any SLA violation may result in renegotiation or moving the job to some other grid service provider.

9.3.2   Grid Infrastructure Solutions

There are two main types of grid infrastructure solution – network solutions and server/host solutions.

9.3.2.1   Grid Network Solutions

There are several solutions which cater to the specific needs of the integration of grid and network technologies like firewalls and VPN. In this section, we will discuss a couple of these, namely the adaptive firewall for grid (AGF) [49], which integrates firewall requirements with the grid computing infrastructure, and hose, which describes flexible resource management using VPN technologies.

Adaptive firewall for the grid (AGF)

The AGF is a project being undertaken at the Technical University of Denmark (DTU). The main motivation behind the work is the observation that to meet the grid firewall requirements, administrators need to open several well-known ports and a range of ephemeral ports for incoming connections. This can be dangerous as attackers may be able to sneak into the system through the open ports. The AGF system develops a mechanism so that the firewall can adaptively open and close ports based on service requests. The firewall will open the ports when it receives authenticated requests. Moreover, it will close them again when there is no service activity on them.

Hose

The hose service model [50] is an effort to provide flexible resource management in a VPN environment. Proposed by researchers from AT&T Research, the hose service model is characterized by aggregate traffic from one set of end-points to another within a VPN. The hose service model is a flexible alternative to the customer-pipe service model, where a customer buys a set of fixed allocations (customer pipes) from the service provider. In this model, the customer specifies the incoming and outgoing traffic aggregated over the different sites in the VPN system. The hose model allows the flexibility of clubbing together traffic with similar QoS requirements. Overall, it provides more flexibility in terms of resource allocation and utilization. This type of model fits nicely with the grid vision as resources can be adjusted on demand. In spite of the flexibility provided by this model, one of its main disadvantages is the lack of QoS guarantees it can provide. Since the resources can be shared, absolute guarantees are hard to provide, which could prevent such a system being accepted widely.

9.3.2.2 Grid Host Solutions

Grid host solutions mostly try to protect the hosts within a grid system through isolation. Solutions like virtualization create multiple virtual machines (VMs) within a single physical machine, resulting in a better isolation environment. Flexible kernels and sandboxing solutions create isolation either through a more protected kernel or trapping system calls. The only type of solution that tries to protect hosts without isolation is the application-level sandboxing solution, which checks the safeness of a code. Proof-carrying code (PCC) is an example of application-level sandboxing.

Application-level sandboxing

Application-level sandboxing is a technique where the isolation and security capabilities are embedded in the application. Security features are hardwired into the application and can be verified before it is executed on a remote system. Cryptographic mechanisms can be used to determine whether a piece of code was produced by a trusted person or compiler. These concepts were used in the development of SPIN kernel [51]. Another seminal work done in the area of application-level sandboxing was PCC [52], developed by George Necula and team from CMU. PCC introduces the concepts of code producer and code consumer. In the former system, the code is produced, and in the latter, the code is executed. PCC is a mechanism by which a code consumer is convinced that the code produced by the code producer is not malicious is nature. To achieve that, the code producer is required to provide a safety proof which guarantees that the code conforms to a formally-defined safety policy. The code consumer then validates the safety proof using a validator to ascertain the safeness of the code.

Virtualization

A typical data center today hosts different applications in different servers, resulting in overprovisioning of resources and low utilization. Therefore, for some time there has been a move toward consolidation of servers to increase the overall utilization of data centers. Research and development in the area of server consolidation has resulted in virtualization solutions in the server-consolidation space. These solutions typically allow applications to run on self-contained environments called virtual machines. It is possible to create different instances of VMs on individual servers, resulting in a better provisioning environment and higher overall utilization. Not only can different instances of VMs run, these instances can also host completely different operating systems (OSs). Therefore, virtualization techniques allow legacy systems to run on new systems seamlessly. In addition to these advantages, virtualization techniques allow the creation of secure environments and can be used as an isolation solution. It is to be noted that the main goal of virtualization solutions is to provide higher resource utilization and server consolidation, and the ability to provide secure and isolated environments is just a by-product of this. Therefore, there is a need to create flexible policies on the virtualized environment. Research is currently being carried out in this regard [53]. To provide virtualization, a layer of software which provides the illusion of a real machine to multiple instances of VMs is necessary. This layer has traditionally been called the virtual machine monitor (VMM). There are also concepts called the host OS and guest OS. The former is the OS which hosts the VMM, and the latter is the OS which is hosted on top of the VMM. It is also possible for the VMM to run directly on the hardware. In that case, host OS is not required, and VMM will play the role of a minimal host OS that runs directly on the hardware. There are three popular virtualization technologies: hosted virtualization, paravirtualization and shared kernel-based virtualization. The hosted virtualization model is one where the VMM and the guest OS run on the user space of the host OS. The applications running on the host OS and the guest OS share the same user space. Generally, this model does not require any modification to the host OS. However, since there are multiple redirections, the performance of such a model suffers significantly. The VMWare GSX server is an example of a hosted virtualization system. The paravirtualization model is one where the OSs are modified and recompiled so that the multiple redirections of the hosted model can be avoided. The performance of the paravirtualization-based systems is better than that of the hosted virtualization-based systems. Xen [54] and Virtuozzo [55] are examples of paravirtualization systems. Shared kernel systems are those systems where the kernel is shared and the user space is partitioned to be used by different sets of applications. An example of a shared kernel-based virtualization system is the Linux VServer [56].

Flexible kernels

Some OS researchers argue that the performance, flexibility and extensibility of OSs are greatly limited by their design, where the interfaces and the implementations of OS abstractions such as interprocess communication and virtual memory are fixed. Flexibility and extensibility were recognized as OS requirements by researchers even in the 1970s [57, 58]. In [57], the authors advocated the design of open OSs, where the system provides a variety of facilities, and the user can use, accept, reject and modify those facilities based on permissions and requirements. In many cases, one facility may become a component on which other facilities are built or developed, like files and disk pages. In that case, there is a need to identify smaller components and make them accessible to the users and other larger components. The development of flexible kernel design provides a whole lot of interesting concepts and designs. We will try to cover the three decades of conceptualization, design and development effort by identifying two representative solutions in this research area. The first, called Hydra [59], was developed in the 1970s, where the researchers separated the policies and mechanisms of the kernel and used them as guiding principles in kernel design. The second, called exokernels [60], developed in MIT, looks at handling resource management at the application layer, thus providing faster transactions and secure operations.

Sandboxing

One of the most popular techniques for achieving isolation is called sandboxing. It is to be noted that many of the solutions mentioned above, such as virtualization, have similar features. However, while the main motivations of virtualization solutions are server consolidation and improving resource utilization, sandboxing solutions were primarily designed with isolation in mind. Sandboxing solutions developed over the years can be broadly divided into three main types: user-level monitoring or system-call trapping, loadable kernel modules and user-level VMs.

9.3.3   Grid Management Solutions

After describing the architecture and infrastructure solutions, let us describe the grid management solutions. There are two types of management solutions: credential-management solutions and trust-management solutions.

9.3.3.1   Grid Credential-Management Solutions

The credential-management systems can be broadly categorized into credential repositories and credential-federation systems. As the name suggests, the *credential repositories* or credential storage systems are concerned with securely storing credentials, generating new credentials on demand and sometimes generating proxy credentials on a user's behalf for delegation purposes. *Credential-federation systems* or credential-share systems are responsible for sharing the credentials across different domains or realms.

Credential repositories

The credential storage systems are designed so that the responsibility of storing the credentials securely is outsourced from the user to these systems, and the user can get the credentials any time on demand. Examples of such repositories are smart cards [61], virtual smart cards [62] and MyProxy Online Credential Repository [63]. The smart card, an intelligent token, is a credit-card-sized plastic card embedded with an integrated circuit chip. It provides not only memory capacity, but computational capability as well. A software form of the smart card was introduced by Sandhu *et al.* [62], where they also distinguished between virtual smart cards and virtual smart tokens. With virtual soft tokens, the user can retrieve the private key in any system of their choice, which is not possible with the virtual smart cards. The MyProxy system was developed in the University of Illinois, Urbana Champagne (UIUC), designed to meet the credential-management requirements of the grid community. The MyProxy toolkit is the grid middleware and is quite popular. It has been used in major grids including NEESgrid, TeraGrid, EU DataGrid and the NASA information power grid. The MyProxy system is the implementation of the virtual soft-token system proposed in [62], where the X.509 proxy certificates are used to store and retrieve user credentials without having to expose the private key. During the enrollment phase, the long-lived user credentials are stored in the MyProxy repository, whose typical lifetime ranges from weeks to years. Users fetch the short-term credentials or proxies (with lifetime set to a week or less) from the MyProxy server so that the long-term credentials are safe. To achieve the above, the client establishes a TCP connection to the server and initiates the TLS handshake protocol, as shown in Figure 9.2. The server must authenticate with the client using its own certificate. The client may also authenticate with the server, but this step is optional for clients who do not possess X.509 credentials. According to Sandhu *et al.* [62], systems similar to the MyProxy system are stronger than physical soft-token systems, but are vulnerable to dictionary attacks. The reason is that the private key is exposed to the server and can be compromised.

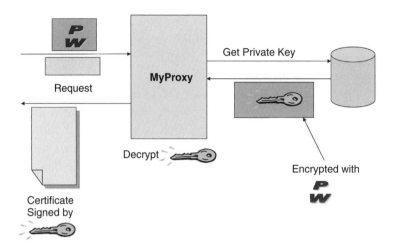

Figure 9.2 MyProxy credential-management system.

Credential-federation systems

These can be broadly categorized into two main types: specific and generic. The specific solutions aim at creating a federated solution for a specific platform or protocol. Examples of such systems are Virtualized Credential Manager (VCMan) [64] and KX.509. The former tries to solve the interoperability issue of CAS by extending CAS to provide interoperability, while KX.509 is a protocol of interoperability between X.509 and Kerberos credentials. Though these solutions solve a niche problem, the generic problem needs to be tackled as well. The Liberty framework [65] from the Liberty Alliance is an attempt in that direction. It is a framework for sharing attribute information in a distributed manner across entities in a trusted domain called the Circle of Trust (COT). Another solution is Shibboleth [66]. Shibboleth is a federated identity-management system based on open-source software developed by the Internet2 consortium members, with assistance from the National Science Foundation. Internet2 is a consortium of United States universities working in partnership with industry and government to develop and deploy advanced network applications and technologies. Shibboleth is essentially a transport mechanism built on top of an institution's existing architecture that allows organizations to exchange information about their users in a secure and privacy-preserving manner.

9.3.3.2 Trust-Management Solutions

Trust-management solutions (TMS) can be divided into two main types: policy-based TMS and reputation-based TMS.

Policy-based TMS

In policy-based systems, the different entities or components constituting the system exchange and manage credentials to establish the trust relationships, based on certain policies. The primary goal of such systems is to enable access control by verifying credentials and restricting access to credentials-based predefined policies. All authorization systems fall under this category. A specific example of policy-based TMS is TrustBuilder [67], which is a trust-negotiation project created through collaboration by researchers in the University of Chicago, Urbana Champagne and Brigham Young University. TrustBuilder develops an infrastructure for trust negotiation on open systems. The TrustBuilder system deploys trust negotiation on TLS, SMTP, POP, ssh and HTTPs. TrustBuilder basically allows strangers to access sensitive data and services over the Internet. In TrustBuilder protocol and architecture, the negotiating parties establish trust between themselves by negotiating trust in a need-to-know manner. In this way, all credentials are not disclosed to either party.

Reputation-based TMS

Reputation-based TMSs provide a mechanism by which a system requesting a resource can evaluate the trust of the system providing the resource. The trust values are a function of the global and local reputations of the systems, along with the different policies. Perhaps the most widely-known TMS is the PeerTrust [68]. This was developed in Georgia Tech with peer-to-peer-based electronic applications in mind. PeerTrust does not use a centralized database for storing trust information. Rather, the trust information is stored in a distributed manner over the network. Each peer or node in the network has a trust manager that is responsible for feedback submission and trust evaluation, a small database that stores a portion of the global trust data and a data locator for placement and location of trust data over the network. Another interesting project that develops a distributed trust and reputation-management architecture is called XenoTrust [69], which is built on the XenoServer Open Platform [70]. The platform was developed at the University of Cambridge. The platform consists of three main components: XenoServer, XenoCorp and XenoServer Information Services (XIS). XenoServers provide services to the client, like hosting client tasks in exchange for money. XenoCorp provides authentication, auditing, charging and payment services. Each XenoServer periodically reports its status and the XIS is used for storing the XenoServer status updates. NICE framework [71], developed at the University of Maryland, is a platform for implementing cooperative applications over the Internet. Cooperative applications can be defined as a set of applications that allocate a subset of resources, typically processing, bandwidth and storage, for use by other nodes or peers in the application. Therefore, grid computing is naturally an application for the NICE trust management framework.

9.3.4   Comments on Grid Solutions

Among the different grid solutions, solutions to the architectural issues are most advanced, and have been deployed in several large grid installations. In enterprises, the issue would be integration, where different enterprise standards, products and protocols need to be integrated with the proposed solutions. The infrastructure solutions and the management solutions are still in the nascent stage and require further research. Table 9.2 provides an overview of the grid solutions.

Table 9.2   Overview of grid solutions.

Category	Solutions	Problems solved	Comments
Grid-architecture-level security issues	GSI	Confidentiality, integrity and authentication	Most of the common information-security problems can be solved using GSI
	Authorization systems	Authentication, authorization	Common authorization problems are solved. However, integration with enterprise solutions like Kerberos is needed
	Grid-services security solutions	QoS violation	These solutions are still research prototypes
Grid-infrastructure solutions	Grid-network solutions	Firewall and VPN integration	These are research prototypes; more research is needed
	Grid-host solutions	Host protection solutions	Virtualization-based solutions have great potential
Grid-management solutions	Credential-management solutions	Protecting credentials	Credential-management solutions have been deployed in the e-sciences grid. Integration is needed for credential repositories and credential federation systems
	Trust-management solutions	Trust management between grid nodes	These are research prototypes and have limited deployment

9.4   Storage-Level Solutions

In Chapter 5, we grouped storage threats and vulnerabilities into two main categories: storage area network (SAN) security and distributed file system (DFS) security.

9.4.1   Fiber-Channel Security Protocol (FC-SP) – Solution for SAN Security

One of the standards that have been proposed for fiber-channel security is the Fiber-Channel Security Protocol (FC-SP) [72]. This standard defines mechanisms that may be used to protect against several classes of threat. These mechanisms include protocols to authenticate fiber-channel entities, protocols to set up session keys, protocols to negotiate parameters to ensure frame-by-frame integrity and confidentiality, and protocols to define and distribute policies across a fiber-channel fabric.

9.4.1.1   Authentication Infrastructure

The fabric-security architecture is defined for several authentication infrastructures. Secret-based, certificate-based and password-based authentication infrastructures are accommodated. Specific APs that directly leverage these three authentication infrastructures are defined. With a secret-based infrastructure, entities within the fabric environment that establish a security relationship share a common secret or centralize the secret administration in a RADIUS server. Entities may mutually authenticate with other entities by using the DH-CHAP protocol. SA may be set up using the session key computed at the end of the DH-CHAP transaction. Frame integrity or confidentiality may be provided by using the ESP_Header. With a certificate-based infrastructure, entities within the fabric environment are certified by a trusted certificate authority. The resulting certificates bind each entity to a public–private key pair, which can be used to mutually authenticate with other certified entities via the Fiber-Channel Authentication Protocol (FCAP). SAs may be set up by using these entity certificates' and associated keys' session keys, computed at the end of the FCAP transaction. Frame integrity or confidentiality may be provided by using the ESP_Header. With a password-based infrastructure, entities within the fabric environment that establish a security relationship have knowledge of the password-based credential material of other entities. Entities may use this credential material to mutually authenticate with other entities using the Fiber-Channel Password Authentication Protocol (FCPAP). SAs may be set up using the session key computed at the end of the FCPAP transaction. Frame integrity or confidentiality may be provided by using the ESP_Header.

9.4.1.2 Integrity and Confidentiality

Integrity and confidentiality are critical components of FC-SP architecture. Frame-by-frame cryptographic integrity and confidentiality, replay protection and traffic-origin authentication are achieved by using the ESP_Header optional header. The ESP_Header processing is performed over selected frames according to a set of traffic selectors maintained in the Security Association Data Base (SADB). Traffic selectors are negotiated when SAs are established. CT_Authentication may be leveraged to provide cryptographic integrity, again through traffic selectors present in SADB.

9.4.1.3 Authorization

Authorization in FC-SP is provided through policies. Two basic types of policy are defined: fabric-wide policies and switch-wide data. Fabric policies may be used to control which switches are allowed in a fabric and which nodes are allowed to connect to a fabric. Policies may be further used to specify topology restrictions within the fabric environment (e.g. which switches may connect to which other switches, or which nodes may connect to which switches). Fabric policies also provide the mechanism for controlling management access to the fabric and the ability to control authentication choices, and to specify security attributes for fabric entities (e.g. nodes and switches). Management access to the fabric may be controlled for common transport or IP access. Switch policies that contain per Switch data, sent to an individual Switch. Zoning policies are defined to encode node-to-node restrictions in a form consistent with the policy model. Policy enforcement occurs whenever a connection is attempted, a management application attempts to access the fabric, or a new policy configuration is activated. The appropriate policy objects are checked to determine whether the requested connection or access is to be allowed or denied. The policy enforcement is carried out locally by the entities involved in the connection or access attempt.

9.4.2 Distributed File System (DFS) Security

The DFS is another critical component of the storage infrastructure. In this section, we will briefly discuss the security features of network file systems (NFSs). We will also discuss network-attached secure disks (NASDs) and OceanStore.

9.4.2.1 NFS Security

Security in NFSs is provided at four levels: identification, authentication, authorization and access control. Identification is the process of communicating or determining the user ID associated with an operation. Authentication is the process of verifying a user. Authorization is the process of determining whether that user is allowed to perform an operation. Access control indicates the methods used to restrict access to an operation. One of the most important elements of providing security in an NFS environment is the storage of user information. There are three different ways of storing this information: local files, Network Information Services (NIS) and Lightweight Directory Access Protocol (LDAP). Client authentication in NFS systems is performed through validation of IP address and through IPSec. Kerberos can also be used.

9.4.2.2 Network-Attached Secure Disks (NASD)

In traditional DFSs, a client wanting to access data must make a request to the file server. The server must then verify the client's authorization and will distribute the file if the appropriate criteria are met. Since the server must interact with every file-access request for every client, this can quickly become a bottleneck. NASD's primary goal is to relieve the server bottleneck by only interacting with each user one time, and providing a 'capability key'. With the capability key, the user can access the appropriate disk(s) directly, without any further server interaction. The disks themselves must be 'intelligent', such that they possess enough internal ability to process capability keys and handle file-access requests directly. There are two servers in the NASD design, one to provide authentication and then the actual file server. NASD does not specify the authentication scheme and recommends using any existing method similar to Kerberos. Upon receipt of authentication, a user sends a request to the file server. The server verifies the authenticity of the request and then provides the user with a capability key that corresponds to the user's rights for file access. After obtaining the capability key, the user can communicate directly with the data disk for all future access requests during a given session. The capability object is the critical aspect pertaining to both the confidentiality and integrity of the system. A file manager agreeing to a client's access request privately sends a capability token and a capability key to the client; together these form a capability object. The token contains the access rights being granted for the request and the key is a MAC consisting of the capabilities and a secret key shared between the file server and the actual disk drive. Clients can then make a direct request to an NASD drive by providing the capability object. The drive uses the secret key, which it shares with the file server, to interpret the capability token, verify the user's access rights and service the request. Since the MAC can only be interpreted using the drive/server-shared secret key, any modifications to the arguments or false arguments will result in a denied request. The novel concept associated with NASD is the placing of part of the data-integrity requirement on the disks themselves. The 'intelligent' disks interpret the capabilities objects, encrypt data and transmit results to clients. To ensure integrity on the client end, the disk uses the same hash MAC combination that allowed it to authorize client access to encrypt and send the data to the client. The client can then verify the integrity of the transmission during the decryption process.

Table 9.3 Overview of storage solutions.

Category	Solutions	Problems solved	Comments
SAN security	FC-SP	Authentication, integrity, confidentiality, authorization	A very comprehensive specification. Solves most of the common problems
DFS security	NFS security	Authentication, confidentiality	Good solution for NFS. Most of the common problems are solved
	NASD	Confidentiality, integrity	Good solution for network-attached devices
	OceanStore security	Confidentiality	Still more research is needed to handle the scale

9.4.2.3 Security in OceanStore

OceanStore [73] uses a large number of untrusted storage devices to store redundant copies of encrypted files and directories in persistent objects. Objects are identified by globally-unique identifiers (GUID), which are generated in a similar fashion to the unique identifiers in the SAN file system (SFS). Each identifier is a hash of the owner's public key and a name. Objects can point to other objects to enable directories. All objects are encrypted by the clients. By replicating the objects among servers, clients can avoid malicious servers deleting their data. The extensive use of replication and public keys makes revocation of access and deletion of data difficult to achieve, but it does provide a nice model for a completely decentralized DFS.

9.4.3 Comments on Storage Solutions

As we have described in this section, FC-SP security specification is very comprehensive and solves most of the common security problems associated with fiber channel. The main problem with SAN is that it has minimal security in the infrastructure, which will result in several of the attacks mentioned in Chapter 5. In DFS, NFS security is quite comprehensive. NASD is a good solution for network-attached storage. More research is required in OceanStore to manage its scale and performance. See Table 9.3.

9.5 Conclusion

Security in infrastructure is critical as it covers the gamut of infrastructure elements. In this chapter, we have looked at three components of IT infrastructure, namely network, middleware and storage. At the network level, we looked at the different solutions pertaining to DoS attacks, DNS attacks and routing attacks, and at information solutions and wireless-network solutions. At the middleware level we focused on solutions pertaining to grid computing. These can be divided into architecture solutions, infrastructure solutions and management solutions. Finally, at the storage level we looked at the FC-SP standard and NASD.

References

[1] Dierks, T. and Allen, C. (1999) The TLS Protocol Version 1.0, RFC 2246, January 1999.
[2] Internet Engineering Task Force (2008) http://www.ietf.org, accessed on June13th, 2008
[3] Stallings, W. (2000) IP security. *Internet Protocol Journal*, **7**(1), 11–26.
[4] Kent, S. and Atkinson, R. (1998) IP Authentication Header, RFC 2402, November 1998.
[5] Kent, S. and Atkinson, R. (1998) IP Encapsulated Security Payload (ESP), RFC 2406, November 1998.
[6] Augustyn, W. and Serbest, Y. (2003) Service Requirements for Layer 2 Provider Provisioned Virtual Private Networks, Internet Draft, draft-augustyn-ppvpn-l2vpn-requirements-02.txt, February 2003.
[7] Rosen, E. and Rekhter, Y. (1999) BGP/MPLS VPNs, RFC 2547, March 1999.
[8] Ould-Brahim, H., Wright, G., Gleeson, B. *et al.* (2002) Network Based IP VPN Architecture Using Virtual Routers, Internet Draft, draft-ietf-ppvpn-vpn-vr-03.txt, July 2002.
[9] Bates, T., Rehkter, Y., Chandra, R. and Katz, D. (2000) Multiprotocol Extensions for BGP-4, RFC 2858, June 2000.
[10] Cisco White Papers (2000) Strategies to Protect against Distributed Denial of Service Attacks (DDoS), February 2000.
[11] Park, K. and Lee, H. (2001) On the Effectiveness of Route-Based Packet Filtering for Distributed DoS Attack Prevention in Power-Law Internet. Proceedings of the SIGCOMM, August 2001, pp. 15–26.
[12] Keromytis, A.D., Misra, V. and Rubenstien, D. (2002) SOS: Secure Overlay Services. Proceedings of the ACM SIGCOMM, August 2002.
[13] Yau, D.K., Lui, J.C.S. and Liang, F. (2002) Defending Against Distributed Denial of Service Attacks with Max-Min Fair Server-Centric Router Throttles. Quality of Service, 2002 Tenth IEEE International Workshop, pp. 35–44.
[14] Allen, J. *et al.* (2000) *State of The Practice: Intrusion Detection Technologies.* Technical Report CMU/SEI-99-TR-028, ESC-99-028, Carnegie Mellon, SEI, January 2000.
[15] Axelsson, S. (2000) *Intrusion Detection Systems: A Survey and Taxonomy.* Technical Report 99-15, Department of Computer Engineering, Chalmers University of Technology,Goteborg, March 2000.
[16] SNORT (2008) http://www.snort.org, accessed on June 13th, 2008.
[17] Kenny, S. and Coghlan, B. (2005) Towards a Grid wide Intrusion Detection System. European Grid Conference, February 2005.
[18] Tolba, M.F., Abdel-Wahab, M.S., Taha, I.A. and Al-Shishtawy, A.M. (2005) *GIDA: Toward Enabling Grid Intrusion Detection System*, CCGrid, May 2005.
[19] http://www.oracle.com/technology/products/bi/odm/pdf/odm_based_intrusion_detection_paper_1205.pdf, accessed on June, 13th 2008.
[20] Ryutov, T., Neumann, C. and Zhou, L. (2005) *Integrated Access Control and Intrusion Detection (IACID) Framework for Secure Grid Computing.* Technical Report, USC, May 2005.
[21] Burch, H. and Cheswick, B. (2000) Tracing Anonymous Packets to their Approximate Sources. Proceedings of the 2000 USENIX LISA Conference, December, pp. 319–27.

[22] Sager, G. (1998) Security Fun with OCxmon and Eflowd. Internet2 Working Group Meeting, November 1998.

[23] Snoeren, A.C., Partridge, C., Sanchez, L.A. *et al.* (2001) Hash-Based IP Traceback. Proceedings of the SIGCOMM, August 2001, pp. 3–14.

[24] Bellovin, S.M. (2000) ICMP Traceback Messages, Internet Draft, March 2000.

[25] Savage, S., Wetherall, D., Karlin, A. and Anderson, T. (2001) Network\support for IP traceback. *IEEE Transaction on Networking*, **1**(3), 226–37.

[26] Xiaodong Song, D. and Perrig, A. (2001) Advanced and Authenticated Marking Schemes for IP Traceback. Proceedings of the INFOCOM, April 2001, pp. 878–86.

[27] Malkin, G. (1998) RIP Version 2, RFC 2453 November, RFC 1058 June, 1988.

[28] (a) Moy, J. (1994) OSPF Version 2, RFC 1583, March 1994. [BGP] (b) Rekhter, Y. and Li, T. (1995) A Border Gateway Protocol 4, RFC 1771, March 1995.

[29] Kent, S., Lynn, C. and Seo, K. (2000i) Secure Border Gateway Protocol (S-BGP). *IEEE JSAC*, **18**(4), 582–92.

[30] Zhang, K. (1998) Efficient Protocols for Signing Routing Messages. Proceedings of SNDSS.

[31] Wang, F., Gong, F., Wu, F.S. and Narayan, R. (1999) Intrusion Detection for Link State Routing Protocol Through Integrated Network Management. Proceedings of the ICCCN, pp. 694–99.

[32] Chakrabarti, A. and Manimaran, G. (2002) *Secure Link State Routing Protocol*. Technical Report, Department of ECPE, Iowa State University.

[33] Smith, B.R., Murthy, S. and Garcia-Luna-Aceves, J.J. (1997) Securing Distance-Vector Routing Protocols. Proceedings of the SNDSS, February 1997, pp. 85–92.

[34] Earle, E.A. (2005) *Wireless Security Handbook*, Auerbach Publications.

[35] Microsoft White Paper, Overview of the WPA Wireless Security Update in Windows XP, 2008.

[36] NIST FIPS 197: Advanced Encryption Standard, http://csrc.nist.gov/publications/fips/fips197/fips-197.pdf

[37] Pearlman, L., Welch, V., Foster, I. *et al.* (2002) A Community Authorization Service for Group Collaboration. Proceedings of the IEEE 3rd International Workshop on Policies for Distributed Systems and Networks, Monterey, CA, pp. 50–59.

[38] Alfieri, R., Cecchini, R., Ciaschini, V. *et al.* (2003) VOMS: An Authorization System for Virtual Organizations. 1st European Across Grids Conference, Santiago e Compostella.

[39] Chakrabarti, A. and Damodaran, A. (2006) *Enterprise Authorization and Licensing Service*. Infosys Technol Report.

[40] Thompson, M., Essiari, A. and Mudumbai, S. (2003) Certificate-based authorization policy in a PKI environment *ACM Transactions on Information and System Security (TISSEC)*, **6**(4), 566–88.

[41] Chadwick, D. and Otenko, O. (2002) The PERMIS X.509 Role Based Privilege Management Infrastructure. ACM SACMAT, Lake Tahoe, CA, pp. 135–40.

[42] Breitbart, Y. *et al.* (2001) Efficiently Monitoring Bandwidth and Latency in IP Networks.Proceedings of the IEEE INFOCOM, Alaska, pp. 933–42.

[43] Chan, M.C., Lin, Y.-J. and Wang, X. (2000) A Scalable Monitoring Approach for Service Level Agreements Validation. Proceedings of the International Conference on Network Protocols (ICNP), Osaka, pp. 37–48.

[44] Dilman, M. and Raz, D. (2001) Efficient Reactive Monitoring. Proceedings of the IEEE INFOCOM, Alaska.

[45] Duffield, N.G., Presti, F.L., Paxson, V. and Towsley, D. (2001) Inferring Link Loss Using Striped Unicast Probes. Proceedings of the IEEE INFOCOM, Alaska, pp. 915–23.

[46] Bradley, K.A., Cheung, S., Puketza, N. *et al.* (1998) Detecting Disruptive Routers: A Distributed Network Monitoring Approach. Symposium on Security and Privacy, Oakland, CA, pp. 115–24.

[47] Barmouta, A. and Buyya, R. (2003) GridBank: A Grid Accounting Services Architecture (GASA) for Distributed Systems Sharing and Integration. International Parallel and Distributed Processing Symposium (IPDPS'03), Nice.

[48] Sandholm, T. (2005) *Service Level Agreement Requirements of an Accounting- Driven Computational Grid*. Technical Report TRITA-NA-05332005, Royal Institute of Technology.

[49] Yao, T.D. (2005) Adaptive Firewalls for the Grid. Master's Thesis, Technical University of Denmark.

[50] Duffield, N.G., Greenberg, P.G., Mishra, P. *et al.* (1999) A flexible model for resource management in virtual private networks, Proceedings of the Conference on Applications, Technologies, Architectures, and Protocols, Computer Communication, ACM Press, pp. 95–108.

[51] Bershad, B., Savage, S., Pardyak, P. *et al.* (1995) Extensibility, Safety, and Performance, in the SPIN Operating System. ACM Symposium On Operating Systems Principles (SOSP), Copper Mountain, CO, pp. 267–83.

[52] Necula, G. (1997) Proof Carrying Code. Principles of Programming Languages, Paris.

[53] VMWare® (2006) http://www.vmware.com, accessed on July 13th, 2006.

[54] Barham, P., Dragovic, B., Fraser, K. *et al.* (2003) Xen and the Art of Virtualization. ACM Proceedings of Symposium On Operating Systems Principles (SOSP), New York, pp. 164–177.

[55] Virtuozzo Team (2005) A Complete Server Virtualization and Automation Solution. Virtuozzo White Paper and Data Sheet.

[56] VServer (2006) http://www.linux-vserver.org/Documentation, accessed on 13th July, 2006.

[57] Lampson, B.W. (1971) On Reliable and Extendable Operating Systems. State of the Art Report, Infotech, 1.

[58] Lampson, B.W. and Sproull, R.F. (1979) An Open Operating System for a Single-User Machine. Proceedings of the Seventh ACM Symposium on Operating Systems Principles, Pacific Grove, CA, pp. 98–105.

[59] Wulf, W., Cohen, E., Corwin, W. *et al.* (1974) HYDRA: the kernel of a multiprocessor operating system. *Communications of the ACM*, **17**(6), 337–44.

[60] Engler, D.R., Kaashoek, M.F. and O'Toole J., Jr. (1995) Exokernel: An Operating System Architecture for Application-Level Resource Management. ACM Symposium On Operating Systems Principles (SOSP), Copper Mountain, CO, pp. 251–66.

[61] Petri, S. (1999) An introduction to smart cards. *Messaging Magazine*, September, Issue, 1–12.

[62] Sandhu, R., Bellare, M. and Ganesan, R. (2002) Password-Enabled PKI: Virtual Smartcards Versus Virtual Soft Tokens. 1st Annual PKI Workshop, pp. 89–96.

[63] Basney, J., Humphrey, M. and Welch, V. (2005) The MyProxy online credential repository. *IEEE Software Practice and Experience*, **35**(9), 801–16.

[64] Mosebach, K., Alves, L.D. and Chakrabarti, A. (2004) Virtualized Credential Management in Inter-domain Grid System. Trusted Internet Workshop (TIW).

[65] Liberty Alliance (2008) Introduction to Liberty Alliance Identity Architecture. Liberty Alliance White Paper and Documentation, available at Liberty Web site, http://www.projectliberty.org/resources/whitepapers/LAP%20Identity%20A   rchitecture%20Whitepaper%20Final.pdf, accessed on June 13th, 2008.

[66] Shibboleth Internet2 project (2008) http://www.shibboleth.internet2.edu, accessed on June 13th, 2008.

[67] Winslett, M., Yu, T., Seamons, K.E. *et al.* (2002) Negotiating trust on the web. *IEEE Internet Computing*, **7**(6), 45–52.

[68] Nejdl, W., Olmedilla, D. and Winslett, M. (2004) PeerTrust: automated trust negotiation for peers on the semantic web, *Proceedings of the Workshop on Secure Data Management in a Connected World (SDM'04)*, Springer,Toronto.

[69] Dragovic, B. and Kotsovinos, E. (2003) XenoTrust: Event-Based Distributed Trust Management. Second International Workshop on Trust and Privacy in Digital Business, Prague (Czech Republic).

[70] Dragovic, B. and Hand, S. (2003) Managing Trust and Reputation in the XenoServer Open Platform. First International Conference on Trust Management, Crete.

[71] Lee, S., Sherwood, R. and Bhattacharjee, B. (2003) Cooperative Peer Groups in NICE. IEEE INFOCOM, San Francisco, CA.

[72] INCITS Working Draft (2006) Fibre Channel Security Protocols (FC-SP), February 2006.

[73] Kubiatowicz, J., Bindel, D., Chen, Y. *et al.* (1999) Oceanstore: An Architecture for Global-Scale Persistent Storage. ASPLOS, December 1999.

10

Application-Level Solutions

10.1 Introduction

Over the last few years there has been a drastic shift in the targets of attacks from networks/hosts to the applications themselves. Today attackers are increasingly concentrating on exploiting the design and coding weaknesses inherent to applications, facilitated by a number of factors such as: the lack of security focus and awareness among software developers, who end up producing defective software; the wide availability of public information about security vulnerabilities and exploits [1, 2]; and the availability of sophisticated free and commercial tools [3, 4] which help in exploiting weaknesses without requiring in-depth security knowledge.

These application security vulnerabilities can be addressed thoroughly only if all the stakeholders in the application development (developers, testers, architects and managers) are aware of the solutions (best design and development practices) available to counter them. Secure application development is possible only through considering security at every stage of the application development life-cycle, right from requirements gathering, through architecture and design, testing and deployment, to maintenance.

This chapter focuses on some of the well-researched and industry-proven security solutions (best practices) which can be incorporated into application design and development by architects and developers. Often there are multiple best practices to counter a given vulnerability, and which one to use depends on several factors such as application architecture, deployment infrastructure and so on. Sometimes it may be desirable to use more than one countermeasure as a defense-in-depth strategy.

Please note that these security best practices are countermeasures to prevent the application security attacks covered in Chapter 6, and do not cover security

Distributed Systems Security   A. Belapurkar, A. Chakrabarti, S. Padmanabhuni, H. Ponnapalli, N. Varadarajan and S. Sundarrajan © 2009 John Wiley & Sons, Ltd

features like firewalls and so on. (Some of the more common security functionality is covered in Chapter 3.) These countermeasures are categorized as per the
application vulnerabilities described in Chapter 6. A comprehensive coverage of
all possible best practices is outside the scope of this book. The interested reader
can refer to [5–7] for more detailed coverage of this subject.

10.2   Application-Level Security Solutions

10.2.1   Input Validation Techniques

Insufficient or no input validation is the key reason for several application security vulnerabilities, such as SQL/LDAP/XPATH injection [8–10], XSS [11] and
buffer-overflow attacks.

Injection vulnerabilities represent a class of vulnerabilities which primarily
result from improper input validation or placing excessive trust on the input
entered by the user. Specifically, the injection vulnerability is one where
user-supplied input is used when executing a command or query. If the
user-supplied data is not properly sanitized and validated, a malicious user can
inject carefully-crafted data to change the semantics of the query or command,
leading to an undesirable consequence. Injection vulnerabilities can have a
severe impact and result in loss of confidentiality/integrity, broken authentication,
arbitrary command execution and, in worst case, complete compromise of
the underlying application or system.

Any user input that is sent back to the user without proper validation results in
XSS attacks. Similarly, user input used without proper range validation leads to
buffer overflow and integer underflow kinds of attack.

Though input validation appears to be a trivial problem, it is in fact a very
involved task to design a proper input validation that takes care of all possible
variations. This is primarily because there is no single definition of what forms
'good' or 'bad' input across applications.

This section describes some of the industry-proven best practices for properly
validating input and preventing injection attacks and similar vulnerabilities.

10.2.1.1   Don't Excessively Trust User Input

Applications receive input from different sources, such as data submitted by users
through browsers using HTTP [12] (this includes both form data and header data
like cookies etc.), databases, files and so on, as well as remote service calls, Web
service invocations and so on. Oftentimes application designers and developers
place excessive trust in the data received from these sources and use it without
sufficient validation. This puts the application in trouble if an attacker has supplied
invalid data. It is critical for applications not to trust data that is coming from a
less trusted source, for example:

(1) Data that is submitted by users through browsers. This never reliable and can be tampered with in many different ways. Developers often consider only the text-entry form fields and don't validate data received from hidden variables, drop-down menus, headers (such as cookie information) and so on.

(2) Data submitted from files. This can contain viruses or other malcrafted data and, if processed directly without proper validation, may cause damage.

(3) Data retrieved from a database which is updated by other applications.

(4) Data submitted over a Web service call in an application-integration scenario. For example, consider a scenario where a tour management company integrates its business application with a car rental company's application. It is critical in such a scenario to authenticate and validate the data that is traveling in both directions between these applications.

It is good practice to consider all input as bad and subject it to strict validation. However, if it is necessary for any other reason to do validation selectively, it is critical to understand the trust boundaries of the application and decide which data to trust.

10.2.1.2   Use Centralized Validation Routines

Developing a well-tested, functionally-rich validation library that meets all your input-validation requirements is very difficult. Oftentimes developers develop validation routines to address specific validation requirements. The disadvantage with this approach is that these validation routines are not well tested and are not flexible when dealing with other, similar data. Validation routines have to stand against several different types of attack vector and it is good practice to centralize the use of all input-validation routines. That way, not only will development and maintenance be easy, input-validation rules will be uniform and consistent across applications. Further, validation routines can be updated at a single place when new vulnerabilities are discovered.

10.2.1.3   Don't Use Client-Side Validations for Security

It is a very common misconception that data subjected to client validations is secure and free from vulnerabilities. Designers and developers often think that data is validated on the client, and that attackers cannot bypass this. The reality is that any client-side validation can be bypassed very easily. It is critical to remember that the server is unaware of the client software used for communication and what it best understands is the specific protocol it supports. It is trivial to bypass client validations like JavaScript in browsers, or to write a standalone client which talks to the server using the same protocol but without any validations. Similarly, there are plenty of proxies and other automated tools that help capture the data communicated over TCP/IP and provide the ability to modify it dynamically online. Client validations exist for better user experience and performance, but never for enhanced security. It is good practice to always subject client input data to validation on the server, even when client validations are performed.

10.2.1.4 Accept Known Good Input and Reject All Known Bad Input

There are several strategies that one can take in validating data. However, from a security standpoint it is a proven best practice to 'allow only known good input' and reject all other input. Considering the numerous ways in which data can be represented, and plenty of ways in which it can be constructed, it is impractical to know what forms a bad input from a security standpoint. Also, techniques keep changing over time, which means that the 'bad data set' has to be continuously updated. The data that is valid in your application context is known to you at the application design time, and hence it is relatively easy to implement 'allow only known good input' effectively. It is good practice to determine what constitutes the right data for your application and validate input against the accepted data rules for type, length, format and range. In some cases, the data that constitutes bad input (e.g. SQL key words as part of login input) is known up front, and for defense-in-depth a strategy to 'reject all known bad input' can be adopted after 'allow only known good input'. Similarly, sanitize the input using an appropriate encoding technique to make it 'safe', especially if you are displaying it back to the user.

10.2.2 Secure Session Management

HTTP is a stateless protocol, which means it is by definition not possible for a Web server to relate requests coming from the same user. To address this issue, the concept of a session is introduced into Web applications, which allows the server to identify the series of requests made by a client as related. Session management is typically carried out by tying the client and server together through some set of identifiers, commonly referred to as session identifiers. Often the session data is stored on the server side and is accessed through a session identifier reference. The client and server exchange the session identifier between them as long as the session is valid. A variety of attacks are possible on applications depending on how the session identifiers are generated, communicated between client and server, and further managed. These include session-hijacking (obtaining a valid session identifier through interception, prediction and brute force) and session-fixation (fixing the session identifier of a genuine user to an attacker-chosen session identifier) attacks. This section discusses some of the best practices for secure session management.

10.2.2.1 Session-Identifier Generation

Session identifiers form a critical link by which the server can relate the requests made by a particular client. Session identifiers are typically stored in cookies set by the server on the client machine (also called session cookies), appended to URLs (as in the HTTP GET method) or assigned to hidden form fields (as in HTTP POST). These session identifiers play a crucial role in session-hijacking attacks, in which the attacker's main objective is to get hold of an active and valid session identifier. The important factor with session-identifier or session-token generation is that they should be very large in size (a huge number set compared to the number of active users at any given point of time, to avoid possible collisions), highly random (uniform distribution without repetition) and unpredictable. Normal pseudorandom-number generators are not suitable for session-identifier generation as they are highly predictable and broken. Use of cryptorandom-number generators is recommended. Cryptorandom-number generators use highly-random sources and cryptographic techniques to come up with good random numbers for session-management purposes. In-house developed session-identifier generation algorithms are often weak as they are only analyzed by a limited number of people and often lack vigorous testing, and their use in applications must be discouraged heavily. Rather, it is good practice to use the session management features provided by the underlying framework, such as J2EE and .NET. Also, it is always a good idea to analyze when possible the way in which the session identifiers are generated, as all session-identifier generation algorithms are not equally secure. Invalidate any session identifiers that existed before a user was authenticated, and create unique session identifiers after authentication.

10.2.2.2 Session-Identifier Communication

It is trivial to sniff a communication over plain HTTP using the abundance of free tools available over the Web. This HTTP traffic includes the session identifiers and once these are known the attacker can launch session-hijacking attacks. To prevent this session-identifier sniffing, it is critical to exchange session identifiers only over encrypted communication channels such as Secure-Sockets Layer (SSL) and Transport-Layer Security (TLS) However, SSL/TLS do not offer enough protection from theft of the session cookie contents if the application is vulnerable to cross-site scripting attacks. So it is generally good practice to encrypt the session cookie contents using cryptography. Use strong crypto-algorithms with sufficient key lengths (e.g. 128 bit). Similarly, for cookie-based sessions, setting the 'SSL only' attribute to true allows the browser to send cookies only on SSL connections, which will improve security.

10.2.2.3 Session-Data Protection

Secure the data stored in-session. Store all session data only on the server side and avoid storing any sensitive data in session cookies on the client side. Also, separate the session authentication information from other data to be stored on the client and place them in separate cookies. Employ appropriate access privileges on the server-side session storage. Carefully consider how the session data is stored on the server; that is, in files or in some sort of a cache database. Care has to be taken when a Web farm is used, as the session data has to be stored separately so that it is available to all servers in the farm. In such cases it is critical to authenticate requests before allowing access to the session data.

10.2.2.4 Session Lifetime

Invalidate sessions without fail after a certain inactive period. The session time-out is often a compromise between user-friendliness and security. However, the session time-out period should be as small as possible. If the session time-out period is considerable, it provides adequate time for an attacker to exploit session-hijacking or session-replay attacks.

10.2.3 Cryptography Use

A general misconception is that security is all about cryptography. While cryptography plays a key role in fulfilling a limited set of security requirements, such as authentication, confidentiality, integrity and nonrepudiation, it does not offer any solution to several other security issues (e.g. denial-of-service attacks).

Though cryptography is a well-researched topic in information security and has several applications, for example secure communications, getting its use right requires experience. Any small error in coding, configuration, algorithm selection or key-length selection, and any wrong assumption, may completely jeopardize the security of the system and the purpose of the cryptography. This section describes some of the best practices for the use of cryptography.

10.2.3.1 Never Develop Custom Algorithms or Rely on Proprietary Algorithms

Developing crypto-algorithms which can withstand varied attacks requires experience, knowledge of security attacks and a high degree of mathematics background. However, developers often implement their own crypto-systems, assuming their implementations are secret and hence secure. 'Security not by obscurity' is the proven principle in the security field and it is not a good idea to rely on the secrecy of an algorithm. Rather, the strength of a crypto-system depends on the security of the keys. The other disadvantage with proprietary algorithms is that they are often not analyzed by many people and are subjected to minimal testing. There are several incidents in the history of information security when proprietary algorithms which were obscured from public scrutiny have been broken and published by cryptanalysts. It is always good practice to use crypto-algorithms that are published and time tested. Never, ever rely on developing custom crypto-algorithms for your purpose.

10.2.3.2 Choose the Right Algorithms and Key Sizes

Mere use of cryptographic algorithms is not going to offer any security to an application. There are symmetric ciphers, asymmetric ciphers, message authentication codes, password-based encryption schemes, key-exchange algorithms, message-digest algorithms and so on, each of which is best suited to address a specific security issue under specific conditions. Also, there are a multitude of algorithms available under each category. All crypto-algorithms are not equally secure. To further complicate the equation, there are different varieties in these categories with specific characteristics. For example, there are stream ciphers and block ciphers under symmetric crypto-algorithms. Stream ciphers are fast and more suitable for protecting information with a short lifetime (e.g. session data), while block ciphers are more suitable for protecting data that has to be secured for a very long duration. Similarly, stream ciphers are more suitable for encrypting data of unknown length (e.g. data coming from some stream) as they allow encryption of the plain data bit by bit, whereas block ciphers are more suitable for encrypting data of a known length, as they require padding at the end if the length of the input data is not an integral multiple of the block size.

10.2.3.3 All Cryptographic Algorithms are Not Equally Secure by Design

There are several crypto-algorithms which have been broken and had their weaknesses publicly exposed. This could be due to an inherent weakness in their design or because of advancement in technology and computing power. For example, Data Encryption Standard (DES) [13], which was a Federal standard for data encryption, is considered broken with current computing power, and is no longer recommended for any serious data-protection purpose. Similarly, MD5 [14], which was once considered a good message-digest algorithm, is no longer a recommended standard as it too has been broken. The SHA-1 message-digest algorithm is considered weakened now, and the recommendation is to go for the SHA 256 or SHA 512 [15] algorithm. For symmetric crypto-algorithms, Advanced Encryption Standard (AES) [16] is the new Federal recommendation.

Each of these crypto-algorithms supports different ranges of key size. The strength of encryption is also a function of key, where a higher key length in general means a higher key space and hence more security against brute-force attacks. For example, 40-bit key lengths for symmetric crypto-algorithms like RC4 [17] are considered weak and broken. Key lengths of the order of 128 bits are a minimum recommendation for symmetric crypto-systems. Similarly, for asymmetric (RSA/DSA) [18] crypto-systems a minimum of 1024 bits is a recommendation, and 512-bit keys are considered broken.

Designers also have to take the regulations and laws of different countries into consideration when choosing encryption algorithms as the use of crypto-algorithms is generally subject to import and export restrictions.

So, it is critical to choose the right algorithms with the right key sizes for the right purpose to ensure the desired security.

10.2.3.4 Choose Secure Random Algorithms

Random numbers are a key requirement for many important cryptographic functions, such as key generation, one-time passwords and so on. However, the normal pseudorandom-number generators are no good for crypto-algorithms, which require secure random-number generators. A secure random-number generator is one which produces highly-random numbers that are difficult to guess and has a large number space with uniform distribution (i.e. the probability of generating any random number in the space is equal). Typically developers use the random-number generators [19] that come as default with their languages (such as Java) for cryptography purposes. However, these are not secure and do not meet the above properties, and developers should choose their random-number-generation algorithms carefully. Also, a real source of randomness is difficult to obtain in computer programs; there are several techniques to improvise this (e.g. mouse movement, process memory pattern etc.). Select the source of randomness carefully. There are several attacks which exploit the weakness in random-number generators to break the algorithms, even though the algorithms are secure and flawless by design. So it is very critical that you carefully choose the correct random-number algorithms and use them in the published way to remain secure.

10.2.3.5 Ensure Secure Key Management

The 'security not by obscurity' principle recommends the use of crypto-algorithms that are publicized and available for public scrutiny, which implies the strength of the algorithm is in the secrecy of the encryption keys used but not in the secrecy of the algorithm itself. This tells you the importance of storing keys securely. Key management dictates the lifecycle of keys, including key generation, key exchange, key storage, key archival and key destruction. Care has to be taken to ensure security in every stage of the key lifecycle. Storing the keys on secure hardware is a recommended best practice, but often developers write keys into file storage on the operating system because of cost considerations. In such cases, it is critical to protect the key files through appropriate access controls so that they are not stolen, or inadvertently accessed or deleted. Similarly, it is a common practice to hardcode the keys into executables, thinking they are safe. This is a very dangerous practice as it is trivial to read the hardcoded keys from binaries and executables using several tools available for free over Internet. It is critical to store the keys in tamper-resistant secure storage and provide secure access to them. It is also a recommended practice to change the keys at regular intervals. There are cryptographic attacks which can exploit the fact that the same keys are used to encrypt different data to determine what those keys are. Similarly, the key archival is a hot target for attackers as it provides a repository of keys at a single place, and hence it is very important for developers to ensure its security.

10.2.3.6 Choose Certified Crypto-Libraries

Mere support for standard crypto-algorithms in libraries is not sufficient to ensure the security of an application, but it is critical that the libraries are from trusted sources without any backdoors and are implemented in a secure manner. Any weakness in the implementation of a crypto-algorithm may completely remove the protection it offers. For example, a perfect implementation of an algorithm with a predictable random-number generator is no good and does not offer any security. It is good practice to use only those crypto-libraries that are certified by reputed security-assurance accreditations like Common Criteria (CC) [20], Federal Information Processing Standards (FIPS) [21] and so on. These standards define strict evaluation criteria from a security point of view for products and crypto-libraries, and provide a high level of assurance about the security of their implementation.

10.2.4   Preventing Cross-Site Scripting

Dynamic HTML pages that write back the user input to the browsers without validating the input can cause cross-site scripting. This vulnerability can be exploited by attackers to inject malicious scripting code instead of valid input, which when written back to the browser is executed with the privileges of the trusted domain, as it is served from the trusted Web site. The user input can come from anywhere that includes query strings, form variables, cookies, headers and so on. Some example applications that are potentially at risk if they construct output from user input that is not validated include:

(1) Search applications that fetch search results based on user inputs.
(2) Bulletin boards and blogs where users enter their input and the same is read by other users.

Best practices or solutions to address cross-site scripting in applications include:

(1) Filtering all user input and removing all special characters from input that includes query strings, cookies, form variables and so on. Special characters are those which might enable a script to be generated, for example < > ; % & ( ) + − and so on. While filtering is an effective technique if you are trying to validate the input at entry, it may not be the best solution in all cases. For example, if your application context demands some of those special characters to be input, filtering them might not be suitable. Though filtering can be done at JavaScript in a client script, it is always best practice

to do it on the server side, as the client scripting can easily be bypassed. One of the disadvantages of the filtering approach is the data loss associated with the filtered characters.

(2) Similarly, filtering the output involves removing any special characters before writing to the client. Care has to be taken while filtering output, otherwise some of the HTML elements may be lost. For example, if the output is to write an HTML tag like <table>, the special characters < > will be filtered. It is good practice to filter the data that is written out, rather the complete HTML.

(3) A safer approach is to encode the data that is received when the application writes it to the browser. Encoding prevents any embedded scripts from executing on the client browser by removing the meaning associated with the special characters. The advantage of the encoding approach is that there is no data loss associated with it. Suggested best practice is to HTML encode or URL encode the data as appropriate while writing to the output.

10.2.5 Error-Handling Best Practices

Applications without appropriate error-handling design often leak valuable sensitive information to attackers, in the form of stack traces that are written to the client and insecure error messages. These may reveal a lot about the specific technologies used, along with version details, database schema details, SQL query strings, login credentials, directory structure and so on, giving the attackers a deeper understanding of the application and enabling them to launch more targeted attacks.

Error messages displayed at failure events should be consistent. It is common for a login page to show different error messages for invalid user ID entry and invalid password entry. This will give big hints to the attacker about whether the user ID or the password is wrong, allowing them to launch more specific attacks. Besides leaking sensitive data, applications that are not designed for proper error handling may remain open in an inconsistent state following a failure event. An application that is failed open into an inconsistent state may be exploited by attackers, granting them unauthorized access to it. So it is very critical that applications should follow the best practices for a secure error-handling design.

Robust error handling also depends on the technologies being used. Some language platforms like Java and .NET support a more structured exception-handling technique, but others, such as C and PHP, do not provide support for exceptions and it is very difficult to cover all possible error conditions in these.

A centralized exception-handling architecture is a good design approach. Best practices for a secure exception-handling mechanism include:

(1) *Fail safe:* applications should fail into a safe state always. Check all fail conditions carefully to see that they fail into a safe and determinant state.

Also review how fatal errors are handled to see that the application will not fail into an indeterminate state.

(2) *Display safe error messages:* evaluate the code to see that all errors are captured and that generic error messages are displayed, rather than stack traces. Also see that error messages are consistent and do not leak any information of help to attackers, such as specific reasons why the error occurred, login credentials, file paths, names and so on. Log detailed error messages with stack traces into appropriate error/event logs and provide alerts to the administrators.

(3) *Use appropriate exception handling where available:* if a structured language with exception-handling support is used, see that a proper exception-handling architecture is in place to catch all exceptions. Catch all exceptions, write the detailed exceptions into event logs and write generic error messages to the consumer. A structured exception-handling mechanism prevents information disclosure. Similarly, if proper exception-handling support is not there in the language and error handling is done through function return values, check for all error return values and handle them appropriately.

10.3 Conclusion

The main causes of application security vulnerabilities are insecure architecture/design decisions and implementation flaws using insecure coding practices. Knowing the root causes for different application security vulnerabilities will help you to build appropriate countermeasures into your application. Chapter 6 covered several known application security vulnerabilities and this chapter discussed some of the industry best practices to avoid them.

References

[1] Web Application Security Consortium Threat Classification, http://www.webappsec.org/projects/threat/classes/cross-site_scripting.shtml.

[2] Open Web Application Security Project (OWASP), http://www.owasp.org.

[3] Source Code Static Analyzers List, http://www.samate.nist.gov/index.php/Source_Code_Security_Analyzers.

[4] Web Application Vulnerability Scanners List, http://www.samate.nist.gov/index.php/Web_Application_Vulnerability_Scanners.

[5] Howard, M. and LeBlanc, D., *Writing Secure Code*, 2nd edition, Microsoft Press, Paperback, Published December 2002.

[6] OWASP, A Guide to Building Secure Web Applications and Web Services 2.0, Black Hat edition, 2005.

[7] Curphey, M. *et al.*, *Improving Web Application Security – Threats and Countermeasures*, Microsoft Press, Published August 2003.

[8] Xpath Injection, http://www.webappsec.org/projects/threat/classes/xpath_injection.shtml, http://www.owasp.org/index.php/XPATH_Injection.

[9] Faust, S. *LDAP Injection, – Are Uour Web Applications Vulnerable?* SPI Dynamics, 2003.

[10] OWASP, SQL Injection, http://www.owasp.org/index.php/SQL_injection.
[11] Spett, K. *Cross Site Scripting, – Are Your Web Applications Vulnerable?* SPI Dynamics, http://spidynamics.com/whitepapers/SPIcross-sitescripting.pdf, 2005.
[12] Hypertext Transfer Protocol (HTTP), RFC2616, http://www.w3.org/Protocols/rfc2616/rfc2616.html.
[13] Data Encryption Standard (DES), http://www.itl.nist.gov/fipspubs/fip46-2.htm.
[14] RFC 1321: The MD5 Message Digest Algorithm, http://www.faqs.org/rfcs/rfc1321.html.
[15] Secure Hash Algorithms, http://www.csrc.nist.gov/publications/fips/fips180-2/fips180-2.pdf.
[16] Advanced Encryption Standard, http://www.csrc.nist.gov/publications/fips/fips197/fips-197.pdf.
[17] RC4 Cipher, http://www.rsa.com/rsalabs/node.asp?id = 2250.
[18] Digital Signature Algorithm, http://www.rsa.com/rsalabs/node.asp?id = 2238.
[19] Cryptographically Secure Random Number Generator, http://www.en.wikipedia.org/wiki/Cryptographically_secure_pseudorandom_number_generator.
[20] Common Criteria for Information Technology Security Evaluation (CC), www.commoncriteriaportal.org/.
[21] Federal Information Processing Standards (FIPS), http://www.itl.nist.gov/fipspubs/.

11

Service-Level Solutions

11.1 Introduction

It is said that if you know both the enemy and yourself, you will fight a hundred battles without danger of defeat; if you are ignorant of the enemy but only know yourself, your chances of winning and losing are equal; if you know neither the enemy nor yourself, you will certainly be defeated in every battle.

Sun Tzu, *The Art of War*

In Chapter 7 we showed how the openness and text-oriented, XML-based nature of service-oriented IT systems have given birth to complex requirements, threats and vulnerabilities. In this chapter we shall concentrate on different solutions to the diverse service-level issues, and mechanisms to handle the threats and vulnerabilities. First, we explore why secure-sockets layer (SSL), the predominant solution for Web-based systems, is not enough for Web services-based systems. Further, we highlight the role of standards in promoting interoperability, a key requirement for service-oriented IT systems. Finally, the emergence of a new breed of firewalls, XML firewalls, is explained, along with their critical role in addressing various service-level threats. We also explore the role of policy-centered security architectures in satisfying key service-oriented security requirements.

11.2 Services Security Policy

The main goal of a corporate security policy is to protect data by defining procedures, guidelines and practices for configuring and managing security in the corporate environment. It is imperative that the policy defines the organization's philosophy and requirements for securing information assets. The information security policy now needs additional considerations, for securing assets that are exposed as services. The security policy should define the minimum security criteria that are required around a service. The policies around services security will

Distributed Systems Security   A. Belapurkar, A. Chakrabarti, S. Padmanabhuni, H. Ponnapalli, N. Varadarajan and S. Sundarrajan © 2009 John Wiley & Sons, Ltd

typically be an extension to the existing information security policy. The security policy is not limited to just network requirements but might additionally define rules and guidelines for design and development of secure Web services.

11.2.1   Threat Classification

The initial step to creating an effective information security policy is evaluating information assets and identifying threats to those assets. Classification of information assets needs to happen based on quantitative (monetary value) and qualitative factors. Determining both the monetary value and the intrinsic value of an asset is essential to accurately gauging its worth. To calculate an asset's monetary value, an organization should consider the impact if that asset's data, networks or systems are compromised in any way. To calculate intrinsic value, the organization should consider a security incident's impact on credibility, reputation and relationships with key stakeholders. When assessing potential threats, both external and internal threats must be considered. Threat classification additionally defines guidelines that classify a service depending upon various parameters, such as:

(1) Classification of information managed by that service.
(2) Internet or intranet deployments.
(3) Underlying host-system classification to the business.
(4) Qualitative and quantitative risk assessment.
(5) Risk probability and assessment matrix.

Examples of the data required for defining a policy around services security are listed below:

(1) Transport-level security requirements around SSL or secure virtual private networks.
(2) Allowed service-communication protocols between the consumer and provider; that is, HTTP, SMTP, SOAP, JMS and so on.
(3) Service proxy or virtualization requirements.
(4) Port numbers to be exposed for the firewall; JMS/MQ could have additional requirements.
(5) Network-address translation or IP hiding requirements.
(6) Authentication requirements like public-key infrastructure (PKI), Kereberos tokens and usernames/passwords.
(7) Single sign-on (SSO) and identity-management requirements.
(8) Authorization requirements like security assertions markup language (SAML) or custom authorization schemes.
(9) Access control requirements to the underlying systems.
(10) Message-level security and validation requirements.

(11) Trusted domains and trusted issuing and requesting authorities.
(12) Other nonfunctional requirements around scalability, availability and so on.

Security is often considered a dead investment and hence is often shortchanged. However, we have also come across scenarios where enterprises have become paranoid and invested a lot of money in security for information which is available in the public domain. Hence the next step is to perform an assessment by weighing security against exposure and the underlying risk. This assessment allows an organization to decide the level of security required for the information. The goal for this risk assessment is to find the optimal line between security costs and risk. This assessment will help in determining the proper allocation of resources once the security policy is effectively in place.

11.3   SOA Security Standards Stack

The standards for security in Web services extend standard mechanisms like encryption, digital signature and public-key infrastructure to handle XML and, in particular, Web services-specific nuances and problems. Some of the protocols are works in progress and some have been standardized already. While some of the standards are quite generic and applicable in generic distributed systems involving XML payloads, specialized standards are only suited for Web services applications. Hence the adoption of some of these protocols is not complete, and in fact there are proprietary protocols implemented by some vendors already. A top-level hierarchy of key SOA standards is presented in Figure 11.1. We shall briefly explore why SSL, the key Web-enabling security standard, is not enough for Web services.

11.3.1   Inadequacy of SSL for Web Services

One of the key factors in the success of the Web as a platform for e-business and e-commerce has been the availability of scalable security algorithms like SSL. SSL has been a key driver in adoption of the Web by enterprises for exposing their functionality. SSL refers to a session-layer protocol in computer networking, whereby a series of exchanges of digital keys happens between a client and a server, ending in a secure channel of communication between the two ends, where messages are sent in encrypted form. SSL-based communication can be of two types:

Server-side SSL requires that the client obtains a copy of the Web server's certificate so that they can authenticate the server. An encrypted channel for data communication is then established between the client and the server. Hence this provides a one-way authentication.

On the other hand, client-side SSL requires both the client and the server to be authenticated, thereby enabling two-way authentication. Hence both service

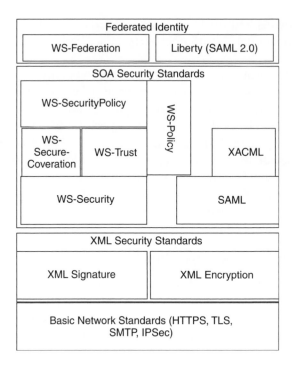

Figure 11.1 SOA standards.

providers and service requesters can be authenticated by using client-side SSL. It might be argued that this is a possible solution for Web services too. But we would say this is not so, for the following reasons:

(1) Web services depend upon the ability of message-processing intermediaries to forward messages, because of the involvement of multihop data transfers, while SSL provides for point-to-point security between two points. On account of this, if we use SSL, the intermediaries will be the weak points where the security breaks down.
(2) Client-side SSL, as discussed above, requires the client digital certificates to be as available as the server digital certificates. The task of distributing client-side SSL certificates to all possible requesters of a Web service is next to impossible.
(3) In a Web application it is usually not necessary to record a transaction as a nonrepudiation proof, but it is important to do so in critical Web services. In this context an SSL session has no memory of an earlier SSL session, and hence fails in providing nonrepudiation. Imagine a Web service which requires only one-time execution of a Web service from a client, denying possibility of a replay attack. This requirement could not be detected with SSL.
(4) Secure sessions between more than two parties is not addressed by SSL.

11.4   Standards in Depth

Standards are crucial to understanding SOA security. We shall therefore explore here the key security standards.

11.4.1   XML Signature

In Web services it is necessary to partially encrypt the body or parts of a Simple Object Access Protocol (SOAP) request, in order to enable transmission with authenticity and integrity assured. Because of the involvement of such multihop data transfers in Web services, the original concept of digital signatures will not extend to XML-based content, as this is based on the idea of getting signatures from the message digests of the entire document. Hence intermediaries need mechanisms for the development of complete trust in the handling of the content of messages keeping partial content intact. Such mechanisms have been provided in the XML Signature [1] specification. XML Signature defines an XML-compliant syntax for representing signatures over Web resources and portions of protocol messages, and procedures for computing and verifying such signatures. These will be able to provide data integrity, authentication and/or nonrepudiation. In real-life scenarios it is necessary that in the transmission route of an XML message, different parties sign different parts. XML Signature specification allows for this kind of signing. XML Signature validation requires that the data object that was signed be accessible. The XML Signature itself will generally indicate the location of the original signed object. This can be referenced by a URI within the XML Signature; it can reside within the same resource as the XML Signature (the signature is a sibling), be embedded within the signature (the signature is the parent – enveloping form) or have the signature embedded within itself (the signature is the child – enveloped form).

Typical computation of the XML Signature of an XML document involves computing the message digest of the document. However, it is necessary to understand that there are many cases where seemingly dissimilar XML documents and nodes actually refer to the same document/node. The idea in canonical XML is to obtain the core of an XML structure, so that any two structurally-equivalent XML documents are identical byte for byte in their core form. This core form is termed the *canonical form* of an XML document. Canonicalization refers essentially to the process of conversion of any XML document to its canonical form, and is necessary for XML Signature computation. An example of an enveloped XML Signature is shown in Figure 11.2. This example uses the DSAwithSHA1 algorithm to compute the signature; the XML Signature provides the key values used to compute it.

11.4.2   XML Encryption

XML Encryption provides end-to-end security for applications that require secure exchange of data. XML-based encryption is the natural way to handle

```
<Signature Id="SriSignature"
      xmlns="http://www.w3.org/2000/09/xmldsig#">
  <SignedInfo >
  <CanonicalizationMethod Algorithm="http://www.w3.org/TR/2001/
        REC-xml-c14n-20010315"/>
  <SignatureMethod Algorithm="http://www.w3.org/2000/09/
        xmldsig#dsa-sha1"/>
    <Reference URI="http://www.w3.org/TR/2000/REC-xhtml1-
          20000126/">
     <Transforms>
      <Transform Algorithm="http://www.w3.org/TR/2001/REC-xml-
          c14n-20010315"/>
     </Transforms>
     <DigestMethod Algorithm="http://www.w3.org/2000/09/
            xmldsig#sha1"/>
      <DigestValue>j71d2Tr3rvEPO0vKtMup4NbeVu8nk=</DigestValue>
     </Reference>
   </SignedInfo>
      <SignatureValue>d12hdlsd7dssf2@4</SignatureValue>
      <KeyInfo>
       <KeyValue>
       <DSAKeyValue>
       <p>33</p><Q>3434</Q><G>3434</G><Y>2323</Y>
       </DSAKeyValue>
       </KeyValue>
      </KeyInfo>
  </Signature>
```

Figure 11.2   Sample XML Signature.

complex requirements for security in data-interchange applications. SSL, as we have already shown, cannot handle encryption of partial messages being exchanged, or secure sessions between more than two parties. XML Encryption is built on top of mature cryptographic technology – in this case, shared key-encryption technology. Core requirements for XML Encryption are that it must be able to encrypt an arbitrarily-sized XML message and it must do so efficiently. Those two factors led its creators to choose shared-key (symmetric) encryption as the foundation for XML Encryption. Encryption provides for message confidentiality (the message will be secret from all but the intended recipient). The reason XML Encryption is needed over and above transport-level encryption such as SSL is that you need to maintain confidentiality of a message in the face of the message taking multiple hops on its way to its destination. This will be common when shared services are utilized. You also need confidentiality when the XML message is stored, even after it reaches its final destination. This requirement is called persistent confidentiality. With XML

Encryption, each party can maintain secure and insecure states with any of the communicating parties. Use of XML Encryption can be broadly divided into three parts:

(1) Encrypting the complete XML document.
(2) Encrypting the XML element.
(3) Encrypting the content of the XML element.

XML Encryption can handle both XML and nonXML (e.g. binary) data. That means we can also send a JPEG file using XML Encryption. Some real-life scenarios where XML Encryption can be applied are:

- Information exchange between two enterprises. One is an online bookseller and the other is a publisher. When the bookseller wants to purchase books, it submits a purchase order to the publisher. At the publisher's end, the sales department receives this order, processes it and forwards it to the accounts department. The two enterprises exchange information in the form of XML documents. Since some portions of the documents need to be secure and the rest can be sent insecurely, XML Encryption is the natural approach for applying security in this case.
- Say we have a secure chat application which has multiple chat rooms with several people each one. XML-encrypted files can be exchanged between chatting parties in such a way that data intended for one particular room will not be easily available to the others.
- Say we need to send an XML file to a publishing company. The file contains details of a book that we need to purchase. It should also contain details about the credit card for the payment. Obviously, we would like to secure this this sensitive data. One option is to use SSL, which secures the whole communication. The alternative is to use XML Encryption. If the application requires that the whole communication be secure, we will use SSL. On the other hand, XML Encryption is the best choice if the application requires a combination of secure and insecure communication (which means that some of the data will be securely exchanged and the rest will be exchanged as is).

11.4.3 Web-Services Security (WS-Security)

WS-Security is the basic standard for message-level security, based on the notion of transmitting relevant tokens as part of SOAP message headers. By adoption of message-level security, the key requirements of SOA security are addressed, primary among them being loose coupling and interoperability. Loose coupling is enabled by separating the information about security from the actual body content. In this context, the typical SOA scenarios involving multiple participants as in a multihop service invocation are addressed by WS-Security headers. The salient features of WS-Security are:

(1) It protects the integrity and confidentiality of a Web-services message.
(2) It provides a mechanism for associating security-related claims with the message.
(3) It improves interoperability, as it resolves some of the ambiguities found in SAML, XML Encryption and XML Signature by standardizing practices.
(4) It provides support for multiple security tokens, multiple trust domains, multiple signature formats and multiple encryption technologies.
(5) WSS 2004 can support both SOAP 1.1 and SOAP 1.2 – it is defined for a generic SOAP specification and is not tied up with any SOAP versions.
(6) It involves concepts of security tokens.
(7) It can leverage existing security mechanisms like username/password, Kerberos, X.509 certificates and so on.
(8) It uses the header in SOAP to store the information.

The Web-services security model is described as follows:

(1) A Web service can require that an incoming message proves a set of claims (e.g. name, key, permission, capability, etc.). If a message arrives without the required claims, the service may ignore or reject it. We refer to the set of required claims and related information as 'policy'.
(2) A requester can send messages with proof of the required claims by associating security tokens with them. Thus, the messages both demand a specific action and prove that their sender has the claim to demand the action.
(3) When a requester does not have the required claims, they or someone on their behalf can try to obtain the necessary claims by contacting other Web services. These other Web services, which we refer to as security token services (STSs), may in turn require their own set of claims. STSs broker trust between different trust domains by issuing security tokens.

This model is illustrated in the Figure 11.3, showing that any requester may also be a service, and any security token service may also be a full Web service, including expressing policy and requiring security tokens.

This general messaging model – claims, policies and security tokens – subsumes and supports several more specific models, such as identity-based security, access-control lists and capabilities-based security. It allows use of existing technologies, such as X.509 public-key certificates, Kerberos shared-secret tickets and even password digests. It also provides an integrating abstraction, allowing systems to build a bridge between different security technologies. The general model is sufficient to construct higher-level key exchange, authentication, authorization, auditing and trust mechanisms.

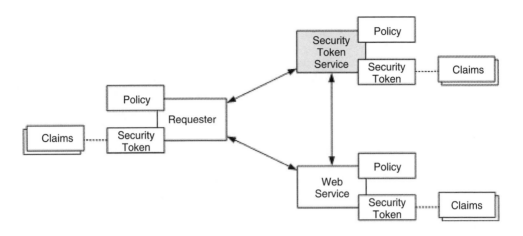

Figure 11.3   Web Services Security model (WS-Security Standard).

11.4.3.1   Tokens in WS-Security

UsernameToken element

A UsernameToken is used when a password is already shared between client and server. Typically at the client, the digest of the password is attached to the SOAP message. This digest contains a hash that is a combination of the password, a nonce (functionally a unique string that identifies a request) and the creation time. At the server, the password hash is created again and compared. Use of timestamp and saving of nonce can avoid replay attacks. An example of a UsernameToken in the WS-Security header of a SOAP message is given in Figure 11.4

BinarySecurityToken element

A BinarySecurityToken provides a means of including X.509 certificates and Kerberos tickets in SOAP security headers. In a typical interaction, authentication is achieved by attaching the public version of a client's certificate to the SOAP message. The message is signed to prove that it has not been tampered with. The signature is generated using the private key.

   A SOAP message with a sample BinarySecurityToken is shown in Figure 11.5. At server, the signature is verified by getting the decrypted hash back using the client's public key and comparing it with the hash of data sent with the message. Encryption can be used for privacy of data. The public key of the receiver is used for encryption and the private key of the receiver for decryption.

SecurityTokenReference element

SecurityTokenReference refers to a means of providing a set of claims that reside at a specific location. Security claims indicate a specific right, privilege

```
<soapenv:Envelope xmlns:soapenv="http://schemas.xmlsoap.org/
    soap/envelope/"
xmlns:xsd="http://www.w3.org/2001/XMLSchema" xmlns:xsi=
    "http://www.w3.org/2001/XMLSchema-instance">
 <soapenv:Header> <wsse:Security soapenv:mustUnderstand="0"
     xmlns:wsse="http://docs.oasis-open.org/wss/2004/01/
     oasis-200401-wss-wssecurity-secext-1.0.xsd">
 <wsse:UsernameToken xmlns:wsse="http://docs.oasis-
     open.org/wss/2004/01/oasis-200401-wss-wssecurity-
     secext-1.0.xsd">
 <wsse:Username xmlns:wsse="http://docs.oasis-
     open.org/wss/2004/01/oasis-200401-wss-wssecurity-
     secext-1.0.xsd">sri</wsse:Username>
 <wsse:Password Type="http://docs.oasis-
     open.org/wss/2004/01/oasis-200401-wss-username-
     token-profile-1.0#PasswordText" xmlns:wsse=
     "http://docs.oasis-open.org/wss/2004/01/oasis-200401-wss-
     wssecurity-secext-1.0.xsd">senthil</wsse:Password>
 </wsse:UsernameToken></wsse:Security> </soapenv:Header>
     <soapenv:Body/></soapenv:Envelope>
```

Figure 11.4 WS-Security header of a SOAP message.

or attribute, for example identity information, keys and so on. Security tokens assert claims that can be attached or referenced.

One of the main drawbacks associated with WS-Security is its use of asymmetric cryptographic algorithms for encryption, which are computationally intensive. To remedy this situation, the WS-Secure Conversation specification was developed, allowing Web services to create a symmetric session key (similar to how SSL/TLS functions) to allow faster symmetric cryptographic algorithms to be used for message-level security. WS-Secure Conversation is well-suited for Web services that receive or send large volumes of messages to a small number of services.

11.4.4 Security Assertions Mark-Up Language (SAML)

SAML is an XML-based framework proposed for exchanging authentication and authorization information among disparate Web-access management and security products. Using SAML, security information can be expressed as an XML document and securely transmitted from one application to another. SAML enables an application to communicate with security systems provided by disparate vendors. Consequently, software from Vendor A can generate information about a user or an access-control decision using SAML, which can be consumed by software from Vendor B without any disclosure of proprietary algorithms or data

```
<soapenv:Envelope xmlns:soapenv="http://schemas.xmlsoap.org/
    soap/envelope/"
xmlns:xsd="http://www.w3.org/2001/XMLSchema" xmlns:xsi=
    "http://www.w3.org/2001/XMLSchema-instance">
<soapenv:Header> <wsse:Security soapenv:mustUnderstand="0"
    xmlns:wsse="http://docs.oasis-open.org/wss/2004/01/oasis-
    200401-wss-wssecurity-secext-.0.xsd">
    <wsse:BinarySecurityToken
EncodingType="http://docs.oasis-open.org/wss/2004/01/oasis-
    200401-wss-soap-message-security-1.0#Base64Binary"
ValueType="http://docs.oasis-open.org/wss/2004/01/oasis-200401-
    wss-x509-token-profile-1.0#X509v3"
xmlns:wsse="http://docs.oasis-open.org/wss/2004/01/oasis-
    200401-wss-wssecurity-secext-1.0.xsd">MIIDAjCCAsACBD/n86swC
    wYHKoZIzjgEAwUAMGcxCzAJBgNVBAYTAlVTMQswCQYDVQQIEwJDQTESMBAG
    A1UEBxMJQ3VwZXJ0aW5vMQwwCgYDVQQKEwNBQkMxEzARBgNVBAsTClB1cmN
    oYXNpbmcxFDASBgNVBAMTC1N1c2FuIEpvbmVzMB4XDTAzMTIyMzA3NTAwM1
    oXDTA0MDMyMjA3NTAwM1owZzELMAkGA1UEBhMCVVMxCzAJBgNVBAgTAkNBM
    RIwEAYDVQQHEwlDdXBlcnRpbm8xDDAKBgNVBAoTA0FCQzETMBEGA1UECxMK
    UHVyY2hhc2luZzEUMBIGA1UEAxMLU3VzYW4gSm9uZXMwggG3MIIBLAYHKoZ
    IzjgEATCCAR8CgYEA/X9TgR11EilS30qcLuzk5/YRt1I870QAwx4/gLZRJm
    lFXUAiUftZPY1Y+r/F9bow9subVWzXgTuAHTRv8mZgt2uZUKWkn5/oBHsQI
    sJPu6nX/rfGG/g7V+fGqKYVDwT7g/bTxR7DAjVUE1oWkTL2dfOuK2HXKu/
    yIgMZndFIAccCFQCXYFCPFSMLzLKSuYKi64QL8Fgc9QKBgQD34aCF1ps93
    su8q1w2uFe5eZSvu/o66oL5V0wLPQeCZ1FZV4661FlP5nEHEIGAtEkWcSPo
    TCgWE7fPCTKMyKbhPBZ6i1R8jSjgo64eK7OmdZFuo38L+iE1YvH7YnoBJDv
    MpPG+qFGQiaiD3+Fa5Z8GkotmXoB7VSVkAUw7/s9JKgOBhAACgYBva6tLG/
    jnzYY8Y0xWF29IFwS9ZXxaMFWROUvPnqqEf3/EOyIeDn9RGEXP0fVW1n/sq
    N6eB11DY95+wF4BrODImrRRnfOFwFcTkMt98TSKJyfc9LYj/MXnYfUHqjKw
    KucPx4gbIWAy3p34iPI3t5buLeaR4NCgivFKW+Sn5j9j+DALBgcqhkjOOAQ
    DBQADLwAwLAIUfBsDEE3xXmQRHboe2jVG0SppaIsCFFOdDJhbfu1IZPRhbQ
    FFz6nMpH6h</wsse:BinarySecurityToken></wsse:Security>
</soapenv:Header> <soapenv:Body/></soapenv:Envelope>
```

Figure 11.5   Sample SOAP message with BinarySecurityToken.

formats. The primary goal of SAML is to provide standardized interoperability between security systems that provide authentication and authorization services. The SAML specification does not define any new technology or approaches for authentication or authorization. Rather, it simply defines a common language for describing the information generated by these systems in XML.

SAML consists of four primary elements:

(1) Security assertions.
(2) Request/response protocol for generating and returning assertions.

(3) Bindings to particular transport protocols (such as SOAP over HTTP).
(4) Profiles for how SAML assertions can be embedded or transported between communicating systems.

An assertion is a declaration of one or more facts (statements) about a subject, for example a user. There are three types of SAML assertion statement, all related to security:

(1) *Authentication statement:* generated in response to an authentication request.
(2) *Attribute statement:* asserts some information about a particular identity.
(3) *Authorization decision statement:* generated in response to an authorization request on a particular resource.

All three statement types have corresponding assertion-request message protocols. For example:

(1) Authentication-query request and response.
(2) Attribute-query request and response.
(3) Authorization-decision query request and response.

All assertions have some common information, such as issuer timestamp, assertion ID and subject. You can extend SAML to make your own kinds of assertion, and assertions can be digitally signed. An example of an assertion with multiple statements is shown in the listing in Figure 11.6.

11.4.5 WS Policy

WS Policy was proposed as a standard to capture generic policy requirements for Web services, to work in conjunction with WS-Security. It:

(1) Provides a flexible and extensible grammar for expressing the capabilities, requirements and general characteristics of entities in an XML Web services-based system.
(2) Defines a framework and a model for the expression of these properties as policies.
(3) Defines a policy to be a collection of policy alternatives, where each policy alternative is a collection of policy assertions. Some policy assertions specify traditional requirements and capabilities that will ultimately manifest on the wire (e.g. authentication scheme, transport protocol selection). Other policy assertions have no wire manifestation, yet are critical to proper service selection and usage (e.g. privacy policy, quality of service (QoS) characteristics).
(4) Provides a single policy grammar to allow both kinds of assertion to be reasoned about in a consistent manner.
(5) Does not specify how policies are discovered or attached to a Web service.

```
<Assertion xmlns="urn:oasis:names:tc:SAML:1.0:assertion"
    AssertionID="2a23f66c-77fb-4e5c-b276-88c8cee06486"
IssueInstant="2003-05-21T11:25:09Z" Issuer="John Doe"
    MajorVersion="1" MinorVersion="0">
<Conditions NotBefore="2003-05-21T11:25:03Z" NotOnOrAfter=
    "2003-05-21T11:28:03Z"></Conditions>
<AuthenticationStatement AuthenticationInstant=
    "2003-05-21T11:25:03Z"
AuthenticationMethod="urn:oasis:names:tc:SAML:1.0:am:
    unspecified">
<Subject><NameIdentifier>johndoe@aol.com</NameIdentifier>
<SubjectConfirmation>
<ConfirmationMethod>urn:oasis:names:tc:SAML:1.0:cm:bearer
    </ConfirmationMethod>
</SubjectConfirmation>
</Subject>
</AuthenticationStatement>
<AttributeStatement xmlns:xsd="http://www.w3.org/2001/
    XMLSchema" xmlns:xsi="http://www.w3.org/2001/XMLSchema-
    instance">
<Subject><NameIdentifier>johndoe@aol.com</NameIdentifier>
    </Subject>
<Attribute AttributeName="name" AttributeNamespace=
    "namespace">
<AttributeValue>Red</AttributeValue>
</Attribute>
</AttributeStatement>
</Assertion>
```

Figure 11.6   Sample SAML assertion.

11.4.6   WS-Trust

Web Services Trust Standard (version 1.3) was proposed in March 2007 (http://
docs.oasis-open.org/ws-sx/ws-trust/200512/ws-trust-1.3-os.html). WS-Trust uses
the WS-Security base mechanisms and defines additional primitives and
extensions for security token exchange to enable the issuance and dissemination
of credentials within different trust domains. In order to secure a communication
between two parties, the parties must exchange security credentials (either
directly or indirectly). However, each party needs to determine whether they can
'trust' the asserted credentials of the other. Some salient features of WS-Trust as
an extension to WS-Security are:

(1) A definition of a means of brokering security credentials among partners
    within different trust domains.
(2) Methods for issuing, renewing and validating security tokens.

(3) Ways to establish, assess the presence of and broker trust relationships.
(4) A flexible set of mechanisms that can be used to support a range of security
    protocols.

The goal of WS-Trust is to enable applications to construct trusted message
exchanges. This trust is represented through the exchange and brokering of security tokens. It provides a protocol-agnostic way to issue, renew and validate these
security tokens. WS-Trust is intended to provide a flexible set of mechanisms that
can be used to support a range of security protocols; this specification intentionally does not describe explicit fixed security protocols. As with every security
protocol, significant efforts must be applied to ensure that specific profiles and
message exchanges constructed using WS-Trust are not vulnerable to attacks (or
at least that the attacks are understood).

At the center of it all is the STS, as shown in Figure 11.3. This is the intermediate broker that issues the security tokens for all incoming requests. WS-Trust
doesn't impose any restrictions on the interactions between the STS and the Web
service with which the requester wants to communicate securely. The STS could
have implicit knowledge of which Web service the requester needs a token for,
or the STS and the Web service might actually be colocated. There could also be
specific instances where there is a dedicated STS which only issues tokens for
communicating with a single Web service.

In most cases a single STS will broker trust for multiple Web services. In these
cases, the request needs to contain some information about the Web service the
requester needs a token for. One way of providing this information is to give the
end-point reference (EPR) of the specific Web service in the request. Requests for
security tokens are made by sending a request security token (RST) message to
the security token service. The current specification defines three possible actions
that can be performed: issuing of a new token, token renewal and token validation.

The following is a sample of a request structure. The request includes an
X509 certificate, which is used as the basis for the request, referred to from
the ws:Base element in the body. The values of the ws:Action header element
and the ws:RequestType element in the body tell the Security Token Service that
the request is for a security token to be issued. The wst:TokenType element in
the body specifies the type of token to be issued, in this case some XML-based
security token. The STS will verify that it recognizes the X509 certificate and
that the holder of the certificate is authorized to be issued with the custom token.

```
<ws:RequestSecurityToken>
<ws:TokenType> mysecuritytoken </ws:TokenType>
 <ws:RequestType>
http://schemas.xmlsoap.org/ws/2004/04/security/trust/Issue
 </ws:RequestType>
 <ws:Base>
 <ws:SecurityTokenReference>
```

```
        <ws:Reference URI='#Sec' ValueType='http://docs.oasis-
            open.org/wss/2004/01/oasis-200401-wss-x509-token-
            profile-1.0#X509v3' /> </ws:SecurityTokenReference>
</ws:Base>
<ws:AppliesTo>
 <ws:EndpointReference>
    <ws:Address>http://security.org/Webservice/ </ws:Address>
 </ws:EndpointReference>
</ws:AppliesTo>
 <wsl:Lifetime>
<wsu:Created>2008-05-06T22:04:34</wsu:Created>
<wsu:Expires>2008-05-08T10:04:34</wsu:Expires>
 </wsl:Lifetime>
</ws:RequestSecurityToken>
```

The following is a sample of a response structure. The wsr:RequestSecurity TokenResponse element (RSTR) is used to return a security token or response to a security token request.

```
<wsr:RequestSecurityTokenResponse>
<wsr:TokenType>
http://schemas.xmlsoap.org/ws/2004/04/security/sc/sct
    </wsr:TokenType>
<wsr:RequestedSecurityToken>
<p:MySecurityToken xmlns:p='http://example.org/mytoken' >
```

Table 11.1 Key fields of a request token.

Element	Description
RequestSecurityToken	Request header
TokenType	Defines the type of security token being requested
RequestType	Indicates to the security-token service the type of action required.
Base	Reference tokens that are to validate the authenticity of the request. For example, could be an existing Kerberos token
Supporting	References the supporting tokens that are needed to authorize the request
AppliesTo	This indicates to the security-token service all the end-point references that the token applies to
EndpointReference	The list of WSDL end-points that are referred to by the requesting service
Lifetime	The validity period or lifetime of a token can be specified, as can information concerning key length, key type and the token issuer
Created	Indicates the time at which a token was created
Expires	Indicates the time at which a token will expire

```
<!-- Token data -->
</p:MySecurityToken>
</wsr:RequestedSecurityToken>
<ws:AppliesTo>
 <ws:EndpointReference>
    <ws:Address>http://security.org/Webservice/ </ws:Address>
 </ws:EndpointReference>
</ws:AppliesTo>
<wsr:Entropy>
<wsr:BinarySecret>XYZ</wsr:BinarySecret>
</wsr:Entropy>
 <wsl:Lifetime>
<wsu:Created>2008-05-06T22:04:34</wsu:Created>
<wsu:Expires>2008-05-08T10:04:34</wsu:Expires>
 </wsl:Lifetime>

</wsr:RequestSecurityTokenResponse>
```

The requester can send a request to the STS to renew the token. A previously-issued token with expiration is presented (and possibly proven) and the same token is returned with new expiration semantics. The STS might also possibly return a new token with completely new expiration semantics.

For this binding, the token to be renewed is identified in the <RenewTarget) element, and the optional <Lifetime) element MAY be specified to request a specified renewal duration. The key semantics around size, type, algorithms, scope and so on cannot be altered during renewal. Token services may use renewal as an opportunity to rekey, so the renewal responses can include a new proof-of-possession token, as well as entropy and key-exchange elements.

The request should have authorized use of the token being renewed, unless there exists a direct trust model between the recipient and the requester to make third-party renewal requests. If intermediary trust models are being used, the third-party requester must prove its identity to the issuer so that appropriate authorization can occur. The renewal binding allows the use of exchanges during the renewal process.

The requester can send a request to the STS to cancel the issued token. When a previously-issued token is no longer needed, the Cancel binding can be used to cancel the token, terminating its use. The request MUST prove authorized use of the token being canceled, unless the recipient trusts the requester to make third-party cancel requests. In such cases, the third-party requester MUST prove its identity to the issuer so that appropriate authorization occurs.

In a cancel request, all key-bearing tokens specified MUST have an associated signature. All nonkey-bearing tokens MUST be signed. Signature confirmation is RECOMMENDED on the closure response. A canceled token is no longer

Table 11.2 Key fields of a response token.

Element	Description
RequestSecurityTokenResponse	Response header to a security-token request
TokenType	Defines the type of security token being requested
RequestedSecurityToken	It is mandatory that at least one wst:RequestedSecurityToken be returned unless there is a negotiation and challenge framework
RequestedAttachedReference	Tokens are considered to be obscure to the requester, and this optional element is specified to indicate how to reference a returned token when it doesn't support references using URI fragments
RequestedProofToken	This optional element is used to return the proof-of-possession token associated with the requested security token. Normally the proof-of-possession token is the contents of this element, but a security-token reference MAY be used instead. The token (or reference) is specified as the contents of this element
AppliesTo	If a wsp:AppliesTo was specified in the request, the same scope needs to be returned in the response by the issuer
Entropy	This element allows an issuer to specify the entropy that is to be used in creating the key
Lifetime	The validity period or lifetime of a token can be specified, as can information concerning key length, key type and the token issuer
Created	Indicates the time at which a token was created
Expires	Indicates the time at which a token will expire
BinarySecret	This optional element specifies a base64 encoded sequence of octets representing the responder's entropy

valid for authentication and authorization usages. On success, a cancel response is returned. This is an RSTR message with the wst:RequestedTokenCanceled element in the body. On failure, a fault is raised. It should be noted that the cancel RSTR is informational. That is, the security token is canceled once the cancel request is processed.

The requester can send a request to the STS to validate the issued token. The validity of the specified security token is evaluated and a result is returned. The result may be a status, a new token or both. For some use cases, a status token is returned indicating the success or failure of the validation. For this binding an applicability scope (e.g. wsp:AppliesTo) need not be specified. It is assumed that the applicability of the validation response relates to the provided information (e.g. security token) as understood by the issuing service.

11.4.7 WS-Security Policy

WS-Security Policy defines a set of assertions for use with the WS Policy framework with respect to security features provided in WSS:SOAP message security, WS-Trust and WS-Secure Conversation. It takes the approach of defining a base set of assertions that describe how messages are to be secured. Flexibility with respect to token types, cryptographic algorithms and mechanisms used, including using transport-level security, is part of the design and allows for evolution over time. The intent is to provide enough information for compatibility and interoperability to be determined by Web-service participants, along with the information necessary to actually enable a participant to engage in a secure exchange of messages. WS Policy defines a framework for allowing Web services to express their constraints and requirements. Such constraints and requirements are expressed as policy assertions.

11.4.8 WS Secure Conversation

Web Services Secure Conversation was proposed in February 2005 (http://specs.xmlsoap.org/ws/2005/02/sc/WS-SecureConversation.pdf).
WS-Secure Conversation defines extensions to WS-Security and builds on WS-Trust to govern secure communication across multiple message exchanges. It defines a mechanism for sharing security contexts and deriving session keys to tie messages together as a part of a 'conversation'.

The security context is defined as a new WS-Security token type that is obtained using a binding of WS-Trust. The mechanisms defined in WS-Security provide the basic mechanisms on top of which secure messaging semantics can be defined for multiple message exchanges. WS-Secure Conversation defines extensions to allow security-context establishment and sharing, and session-key derivation. This allows contexts to be established, and potentially more efficient keys or new key material to be exchanged, thereby increasing the overall performance and security of the subsequent exchanges.

11.4.9 XKMS (XML Key Management Specification)

XKMS stands for XML key management specification and consists of two parts: XML key information service specification (XKISS) and XML key registration service specification (XKRSS). XKISS defines a protocol for resolving or validating public keys contained in signed and encrypted XML documents, while

XKRSS defines a protocol for public-key registration, revocation and recovery. The key aspect of XKMS is that it serves as a protocol specification between an XKMS client and an XKMS server, in which the XKMS server provides trust services to its clients (in the form of Web services) by performing various PKI operations, such as public-key validation, registration, recovery and revocation on their behalf. Now let's talk about why we need XKMS. One of the obstacles to PKI's wide adoption is that PKI operations such as public-key validation, registration, recovery and revocation are complex and require large amounts of computing resources, which prevents some applications and small devices such as cell phones from participating in PKI-based e-commerce or Web-services transactions. XKMS enables an XKMS server to perform these PKI operations. In other words, applications and small devices, by sending XKMS messages over SOAP, can ask the XKMS server to perform the PKI operations. In this regard, the XKMS server provides trust services to its clients in the form of Web services. XKMS defines a Web-services interface to PKI. This makes it easy for applications to interface with key-related services, like registration and revocation, and location and validation. Most developers will only ever need to worry about implementing XKMS clients. XKMS server components are mostly implemented by providers of PKI providers, such as Entrust, Baltimore and VeriSign. VeriSign, for example, provides an XKMS responder that can be used to register and query VeriSign's certificate store. Even SSL server IDs can be validated in real time using the XKMS interface.

XKMS is a foundational specification for secure Web services, enabling Web services to register and manage cryptographic keys used for digital signatures and encryption. When combined with WS-Security, XKMS makes it easier than ever for developers to deploy enterprise applications in the form of secure Web services available to business partners beyond the firewall. XKMS can be used by developers to integrate authentication, digital signature and encryption services, such as certificate processing and revocation status checking, into applications in a matter of hours – without the constraints and complications associated with proprietary PKI software toolkits.

11.4.10 WS Privacy and P3P

WS Privacy is a specification that describes the model for how a privacy language can be embedded into policies. WS-Security is used to associate privacy claims with a particular message. WS-Trust can be used to evaluate these claims for user preferences or organizational policies. WS Privacy is currently being addressed under the P3P (platform for privacy preferences) project. P3P currently addresses Web sites to extend their ability to express privacy preferences in a standard format that can be retrieved implicitly and implemented by user agents. P3P user agents will allow users to be informed of site practices (in both machine- and human-readable formats) and to automate decision-making based on these practices when appropriate. This allows a user to make an informed decision about the Web sites they visit, and also removes the impediment of having to read the privacy policy of each Web site. One of the guiding principles of P3P is to maximize privacy, user confidence and trust on the Web. The P3P [2] specification defines:

(1) A standard schema for data a Web site might wish to collect, known as the P3P base data schema.
(2) A standard set of uses, recipients, data categories and other privacy disclosures.
(3) An XML format for expressing a privacy policy.
(4) A means of associating privacy policies with Web pages or sites, and cookies.
(5) A mechanism for transporting P3P policies over HTTP.

The same philosophy can be extended to Web services, where service-consumer agents can dynamically use these policies to determine if the privacy policies meet their standards and make a decision at runtime regarding the consumption of the service.

P3P policies use XML constructs with namespaces to provide contact information for the legal entity making the representation of privacy practices in a policy. They also explain the date elements that could possibly be captured and how the data could potentially be used. In addition, policies identify the data recipients and make a variety of other disclosures, including information about dispute resolution and the address of a site's human-readable privacy policy. P3P policies must cover all relevant data elements and practices. The elements of the P3P policy vocabulary are given in Table 11.3.

Figures 11.7 and 11.8 indicate the various possible forms of communication between a Web-service consumer and a Web-service provider.

Figure 11.7 represents the scenario where a service consumer requests the privacy policies of the Web-service provider. The request might contain the acceptable privacy policies for the service requester in the SOAP request headers. Upon

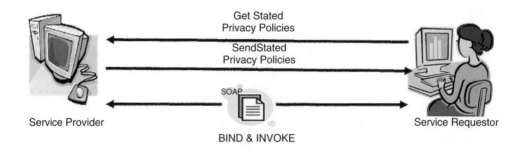

Figure 11.7 Implementing privacy among Web services.

Table 11.3 P3P vocabulary.

Element	Description
ENTITY	Gives a precise description of the legal entity making the representation of the privacy practices
ACCESS	Indicates whether the site provides access to various kinds of information
DISPUTES	Describes dispute resolution procedures that may be followed for disputes about the services' privacy practices, or in cases of protocol violation
REMEDIES	Specifies the possible remedies when a policy breach occurs
STATEMENT	May state that there is no data collected under this <STATEMENT>, or that all of the data referenced by that <STATEMENT> will be kept anonymous upon collection
CONSEQUENCE	Explains why the suggested practice may be valuable in a particular instance
PURPOSE	Must contain one or more purposes for data collection. Must explicitly state the possible purposes of data collection, such as telemarketing, administration and so on.
RECIPIENT	Indicates the legal entity beyond the stated service provider to whom the data could possibly be distributed, for example business partners and so on.
RETENTION	States the type of retention policy. This policy indicates whether the data will be retained for a specific time frame or indefinitely
CATEGORIES	Provide a small insight into the potential uses of data. Could indicate demographics, surveys and so on.

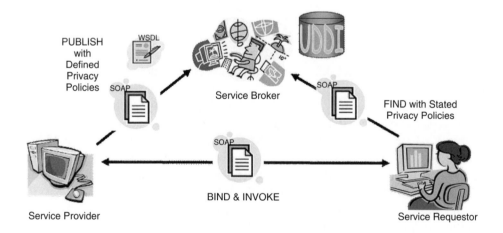

Figure 11.8 Implementing privacy among Web services with brokers.

receiving this request, the Web-service provider sends across the published privacy policies to the service requester. The service requester then analyzes the privacy policies to check if they are secure enough to satisfy their privacy needs. If they are satisfied with the provider's privacy policies then they send across the normal SOAP request.

Figure 11.8 is an alternative approach, where a Web-service broker performs the negotiations between the service provider and service requester. The service provider publishes all their services with their privacy policies for each Web service or even each method published in the Web service. The service requester sends a request, which contains the parameters of privacy that are acceptable to them. The service broker then analyzes this request and matches it against the services published in the registry. The service broker additionally matches the privacy parameters of the service requester against the published privacy policies of the services. After this matching, the service broker returns a set of service end-points that satisfy the request parameters and match the acceptable privacy parameters set by the service requester.

11.4.11   Federated Identity Standards – Liberty Alliance Project and WS Federation

The two crucial standards enabling federated identity in the context of SOA are the Liberty Project (now merged with SAML 2.0) and WS Federation (as part of WS-Security stack). See Table 11.4.

11.4.12   WS-I Basic Security Profile

The Basic Security Profile (version 1.0) (BSP 1.0) was published in March 2007 by the Web Services Interoperability Organization (WS-I). The WS-I Profiles are guidelines that establish tests to ensure secure interoperability between Web services from multiple vendors. The Basic Security Profile is a superset of the WS-I Basic Profile, and defines how to implement Web Service Security in a way that will be interoperable with other Web services that implement the same profile.

The WS-I Basic Security Profile is an interoperability profile that addresses transport security, SOAP messaging security and other security considerations for the WS-I Basic Profile. It defines the interoperability requirements for implementing HTTP over TLS (a point-to-point technology), and the requirements for implementing SOAP Message Security, which provides security protection for SOAP messages even across multiple intermediaries. The WS-I Basic Profile and Basic Security Profile are guides on how to implement the Web-service specifications to improve interoperability.

Table 11.4   Comparison of WS Federation and Liberty Specs.

	Feature/functionality	Liberty alliance project	WS Federation
Similarities	Client profiles	Specifies client profiles for both browser and smart clients	Specifies client profiles for both browser and smart clients
	SSO control flows	SSO control flows specify both front- and back-channel mechanisms	Recommends the front-channel mechanism and discourages use of 'pointer-based' back-channel mechanisms
Differences	Account federation	Opaque identifiers used in identity mapping	Pseudonym service used in identity mapping
	Privacy	Recommends access control policies, usage directives and pseudonymity	Optional privacy support
	Security tokens	Uses SAML and WS-Security	Uses WS-Security
	Business and policy issues	Has business guidelines and authentication context	Business trust issues not addressed
	Underlying technology	Builds on SAML and relies on SSL and WS-Security for transport and message security	Builds on WS-Trust, WS Policy and WS Metadata foundation and relies on SSL and WS-Security for transport and message security
General Differences	Approach	Developed by an open standards community, which includes vendors, end-users and nonprofit organizations	Developed by Microsoft, IBM, VeriSign, BEA and RSA Security; submitted to OASIS
	Scope	Holistic focus on technology, business and policy issues associated with federated identity services	Focus on technology specifications for federated identity services

11.4.13   Status of Standards

Table 11.5   Web-services security standards.

Security standard	Governing organization/status	Comments
XML Signature	W3C Recommendation	Widely accepted. Provides for digital signing of XML documents
XML Encryption	W3C Recommendation	Widely accepted. Provides for encrypting all or part of an XML document
XML Key Management Specification	W3C and IETF Recommendation	Newer. Used in conjunction with XML Signature and XML Encryption. Provides a standard for distributing and registering public keys
WS-Security	OASIS Standard	Widely accepted. Significant tool and vendor support. Standardizes how security is added to SOAP messages
WS-Trust	OASIS Draft	Not yet adopted by OASIS, yet significant vendor support. Extends WS-Security to address interoperability between security tokens
WS-Secure Conversation	OASIS Draft	Builds on WS-Security and WS-Trust. Support from newer tools. Defines how to exchange multiple secure messages as part of a secure conversation
WS Federation	OASIS Draft	Widely accepted even though it is not yet a standard. Recent significant tool and vendor support. Provides for secure access of resources by federated identities
Liberty	Project Liberty	Widely adopted. Now merged into SAML 2.0
WS-Security Policy	OASIS Draft	Part of WS-Security Stack for incorporating security policies
WS Policy	W3C Recommendation	Fast-tracked at W3C. Fills gap in SOA, allowing services to specify their capabilities and security-policy requirements
WS-I Basic Security Profile	WS-I Organization	Newly published by the WS-I to address interoperability of WS-Security standards. Will be widely adopted because it defines how to implement a security model that will interoperate with other security implementations
SAML	OASIS Standard	Widely adopted. Large vendor support. Provides a way to specify authentication information about a user
WS Privacy	P3P project	Wide adoption in Web sites
XACML	OASIS Standard	Newer, yet already widely accepted. Provides a syntax for authorization information

11.5   Deployment Architectures for SOA Security

Key to enforcing standards to increase their effectiveness in achieving SOA security is the right architectural prescription. In terms of SOA security requirements, there are broadly two main architectural components:

(1) Message-level security and policy infrastructure.
(2) XML firewall.

11.5.1   Message-Level Security and Policy Infrastructure

As mentioned in Chapter 7, it is imperative that deployment architectures for SOA should enable loosely-coupled mechanisms for addressing diverse identities, service providers and directories, while ensuring the right security imperatives. In view of such requirements, an equally important imperative has emerged from our discussion of SOA security standards in the previous section, namely that message-level security infrastructure is vital for correct deployment of SOA standards. Likewise, given the diverse policy formats and requirements that need to be accommodated in managing SOA security, it is important that a loosely-coupled, scalable SOA security architecture is put in place, wherein the key drivers of interoperability, plugability and flexibility are easily achievable.

A scalable architecture involving policy infrastructure and capable of handling Web-service authorization and authentication requirements can be illustrated as in Figure 11.9.

This architecture essentially decouples any policy or security constraints from the coding of the business logic of the system. At the same time, the architecture is able to manage multiple existing mechanisms, be they existing LDAP/ID stores or existing authentication mechanisms. The XACML policy engine used in this architecture provides the policy decision and enforcement points. XACML allows authorization policies to be defined and enforced against incoming SAML assertion requests.

The key here is to note that SAML and WS-Security standards are being used interchangeably. While this represents the most generic of the security architectures for SOA, minor variations get induced depending upon the complexity of the underlying security requirements.

11.5.2   XML Firewalls

As mentioned earlier, XML firewalls are the new breed of firewalls, which operate at or above the application layer in the conventional TCP/IP stack. Traditional packet-filtering firewalls are incapable of handling denial-of-service (DOS) attacks at the XML and Web services layer. The XML and Web services-layer attacks are based on text fragments of XML, which cannot be diagnosed or detected by packet-filtering mechanisms.

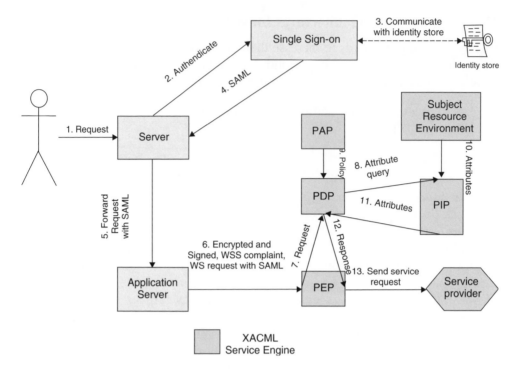

Figure 11.9 Reference security architecture for WS-Security.

In view of these restrictions, the XML firewalls, based on the idea of XML content inspection, are emerging. XML firewalls are available as software proxies or appliances. The software firewalls are generally high-performance proxies that provide additional defense to secure Web service deployments. These XML firewalls can be configured to perform authentication, authorization and auditing, and prevent the various Web-service attacks. They are typically deployed behind the IP firewall, and control the message flow before it reaches the Web services. XML firewall appliances rank higher in terms of performance and scalability, however.

XML firewalls inspect incoming and outgoing messages for vulnerabilities before the messages pass on to an application or client. They are designed to address familiar Web-based attacks that can be transported via XML. Unlike a traditional firewall, an XML firewall works on a higher level and decides whether or not an XML message may access a specific operation of a specific Web service. The access is typically defined in terms of service-requester role and role-based access control (RBAC) rules. Correct authentication allows access to the specific operation and/or data item.

Software XML firewalls are cheaper and have a lower total cost of ownership, but consequently provide less features. XML firewall appliances, on the other hand, can provide better performance and scalability. The general recommendation is always inclined toward the hardware-based XML firewall appliances due to the evolving nature of Web-service standards and the fact that the overall cost is not that prohibitive for an enterprise landscape. A research prototype of an XML firewall, to help prevent XDOS attacks, is available at [3].

11.5.2.1 XML Firewall Request Processing Design

XML firewalls are generally deployed at the edge of the enterprise-application ecosystems. All XML requests are routed through the XML firewall. Hence the XML firewall acts as a gatekeeper and security policies can be defined that determine the different actions taken during the processing of a request. The following is a general list of the steps taken during such processing:

(1) Parse the message headers for the following:
   (a) check source IP address
   (b) check requested resource (physical address/virtual address).
(2) Log the request or send the request to a real-time intrusion-detection system if that has been configured on the firewall.
(3) Check the underlying SSL certificate for its validity if the transport used is HTTPS.
(4) Classify the request based on URL, IP address and the SOAP headers.
(5) If a SOAP action header exists then associate the request with the action handler.
(6) Parse the SOAP envelope for credential checks.
(7) Apply the appropriate security-realm handler for the following credentials:
   (a) username/password handler
   (b) WS username token handler
   (c) Kerberos token handler
   (d) SAML token handler
   (e) Any other custom SSO token-management handler.
(8) If the credential check is successful then associate a message handler with the message, depending on the message classification. There may be a chain of handlers that can be associated in a configured order.
(9) Check message for well-formedness and schema conformance.
(10) Evaluate the SOAP body of the message and make any additional credential checks that may be required.
(11) Message-level encryption could potentially exist, so possibly configure the firewall to check for the checksum validity to discover any message tampering.
(12) Decrypt the XML content and validate the XML Signature if associated with the message.

(13) Possibly configure the firewall to check for message complexity and size, to manage DOS attacks.
(14) Possibly apply XSLT validations to the message body to validate against the schema.
(15) Possibly apply constraint-based filters to the messages, to check the validity of the message values.
(16) Possibly apply pattern-matching filters to check for injection attacks.
(17) Check for any SOAP attachments associated with the message. Additional parameters can be configured to check for the attachment size and viruses, to prevent any hidden attacks.
(18) Process any additional SOAP headers if required.
(19) Determine the routing, depending upon whether the access requested is to:
    (a) a single physical service
    (b) a single virtual service
    (c) multiple virtual services.
(20) Apply any postprocessing XSLT to the message.
(21) Recheck message validity, and generate a timestamp and audit the message.
(22) Route the message to the destination.

11.5.2.2 XML Firewall Deployment Architectures

The two most common deployment scenarios for the XML Gateway are detailed below.

Demilitarized zone scenario

The demilitarized zone (DMZ) scenario (Figure 11.10) is a common deployment design for the perimeter security of applications. In a DMZ scenario two IP firewalls are used to create a demilitarized zone. The XML security gateway resides in this DMZ and sniffs and analyzes all incoming XML and SOAP traffic. The Web service end-point lies at the other end in a protected network. The whole design is based on a 'defense of depth' mechanism. The exterior IP firewall controls, restricts and enforces policies regarding access to the public network. The XML security gateway is used to perform authentication and authorization on the XML traffic. The interior firewall is responsible for monitoring traffic to the protected network. There is also an additional XML security gateway that monitors and controls XML traffic originating from the protected network or the internal users of an enterprise.

Federated trust (B2B) scenario

The second scenario (Figure 11.11) is a more advanced deployment design used for integrating business partners with the enterprise. A prime example is manufacturers integrating their suppliers into their lifecycle for 'just in time' production.

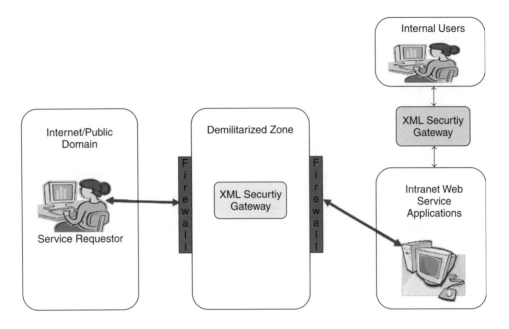

Figure 11.10   DMZ deployment scenario.

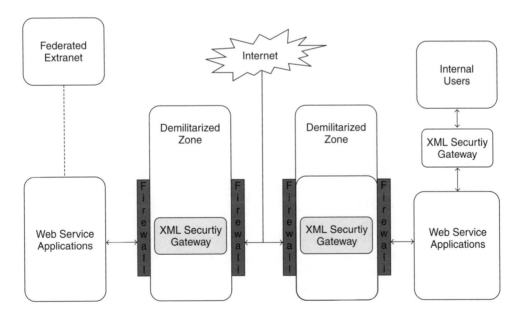

Figure 11.11   Federated deployment scenario.

This is the virtual integration of heterogeneous systems across multiple organizations with federated users. While this allows for greater flexibility, it also gives attackers the opportunity to break into systems across different enterprises. It also means that an enterprise is only as secure as the weakest link of the federated network. The design is literally a double mirror image of the DMZ scenario, connected by the Internet. The underlying network connections could be secure private networks as per the security needs and the criticality of the information flow. XML gateways are installed on the service requester and the service provider. The XML gateway on the service requester provides access control and can enforce policies to control access to services and to the information in the requests. The gateways between the service provider and service requester need to be interoperable on user identity and their underlying roles, and probably even to provide SSO. The service gateways may use SAML as the common framework for authentication, authorization and even SSO token exchanges.

11.6   Managing Service-Level Threats

An XML gateway or firewall generally provides the ability to create, configure and enforce a security policy. As all XML traffic is directed through the firewall, it becomes a central point at which to enforce security policies. We shall now list the solutions that should be offered by XML firewalls to combat the security threats listed in Chapter 7. We use a set of logical parameters for the configuration of an XML firewall, to help us address individual attacks, and these parameters may exist in different forms in different frameworks or products.

Services threat management is an integral part of the overall design of the security architecture. The services security design should consider the following:

(1) Data encryption.
(2) XML packet inspection.
(3) Easy and quick definition and implementation of security policies.
(4) Message inspection, validation and integrity checks.
(5) Support for various authentication mechanisms.
(6) Authorization and access control.
(7) Service virtualization.
(8) Real-time check for digital certificate validations.
(9) Support for digital signatures for signing transactions.
(10) Support for federated trust models.
(11) Network-perimeter defense design.
(12) Transport-level security.
(13) Real-time fraud management.
(14) High levels of security logging and auditing.

11.6.1   Combating SQL and XPath Injection Attacks

There are various techniques that can be used to prevent code-injection attacks. Some of these are listed below.

11.6.1.1   Malicious Input Validation

Input validation is the most important step in avoiding injection attacks. All input is guilty until proven otherwise. All input should be constrained with a set of values. It should be validated and sanitized before it is processed by the service. Validate not only the type of data but also its format, length, range and contents (for example, a simple regular expression such as if $(/^" *^ ' ; \& < > ()/)$ should find most suspect special characters). Validate data both on the client and on the server, because client validation is extremely easy to circumvent.

11.6.1.2   XML Firewall Filters

Injection can be prevented using XML firewalls. XML firewalls support the configuration of pattern-matching filters. The filters allow checking for specific patterns and can replace single apostrophes with double apostrophes. This will force the database server to recognize the apostrophe as literal characters rather than as string delimiters.

11.6.1.3   XSL Validators

Additionally, the incoming XML message inputs can be transformed using XSLs. The XSLs can perform validations on specific parts of the input messages. This ensures the validity of the input.

11.6.1.4   Avoid Dynamic Queries

Avoid the use of dynamically-generated queries. For example, using the Prepared-Statement interface as opposed to the Statement interface in the java.sql package is advocated. The use of parameterized query statements or stored procedures will make it extremely difficult to mount injection attacks.

11.6.1.5   Error Information Sanitization

The error or SOAP fault information should be sanitized before sending it across to the service requester. There have been so many instances where underlying database errors are passed to the consumer. These can be used by attackers to gain valuable information about the database.

11.6.2   Combating Cross-Site Scripting Attacks

Phishing attacks, or attacks where what appears to be a valid URL links to a fraudulent Web page whose purpose is to collect users' data, are nothing new to the Web world. Most of the problems related to cross-site scripting are caused by existing flaws in the security designs of applications, and simple mistakes committed by users. The following techniques can be used to avoid XSS attacks.

11.6.2.1   Filter Input Parameters

It is important to filter all special characters from input that is stored or processed on a server because URLs and GET/POST requests can be created manually. Special characters are characters that enable script to be generated within an HTML stream.

11.6.2.2   Encode Output Based on Input Parameters for Special Characters

It is important to encode the received input data before it is written down to the HTML stream. This technique is effective on data that was not validated for some reason during input. It also encodes URLs, making it difficult for attackers to inject malicious JavaScript. However, given that eventually the application is responsible for all of the UI or response streams, it is imperative that systems be designed to validate all inputs provided by the user.

Users are always advised to look carefully at URLs and the parameters provided with them. URLs will appear in the status bar of your browser and you should always look for external script reference.

11.6.3   Combating Phishing and Routing Attacks

Some common techniques to address Schema-based attacks, WSDL phishing, probing attacks and SOAP routing attacks are included below.

11.6.3.1   Service Virtualization

WSDL is the primary source of information for mounting attacks. Hence it is imperative to protect the WSDLs. XML firewalls can be used to virtualize the underlying URLs and services. Virtualization allows the provider to hide URL and the communication protocols from attackers.

Service virtualization can be implemented as an intermediary or through WSDL proxies. An additional layer of security could be to implement message-level filters. These filters can be used to perform content based routing to an internal name or URL. It is also advisable to enforce authentication even for WSDL access.

11.6.3.2   Use of WS Addressing

WSDL end-points used to be located through conventional URLs. This allowed attackers to directly access Web services using a URL such as http://soap.amazon. com/schemas2/AmazonWebServices.wsdl.

WS Addressing has introduced the concept of an EPR. An EPR allows a service requester to request a resource. The service provider can then validate their credentials even before the WSDL has been requested. This prevents an attacker from gaining direct access to the WSDL. Additionally, because EPRs allows for user-defined properties to be sent to the resource, they can actually be bound to different transport protocols. Multiple EPRs might share the same URI but specify different resource reference properties and hence represent different services. This is a behavior akin to polymorphism in objects-oriented languages, where objects can take up different forms at run-time because of some specific properties. This means the attacker sees the same URI for all services and requires further knowledge of resource IDs. Resource IDs can be distributed to concerned partners and consumers separately as passwords. This prevents random WSDL scanning and port scanning.

11.6.3.3   IP-Based Access Control

XML firewalls can be configured to only allow requests from a list of trusted and secure IPs. This prevents unauthorized and attacker requests from coming into the application zone.

11.6.4   Handling Authentication Attacks

11.6.4.1   Handling Dictionary/Forced-Entry Attacks

Dictionary attacks are mounted when an attacker targets a Web service with tools and automated password generators. Cryptography techniques and transport-level protocols like SSL/TLS, in conjunction with Web-services security standards including XML Encryption and WS-Security, can be used to address these. Secure transport protocols (SSL/TLS) can assure the security of messages only at socket-layer security, encrypting all communication over a particular TCP connection. As already discussed, SSL is a good solution for server-to-server security but it can't adequately address a scenario where a SOAP request is routed via more than one server, or only part of a message needs encryption. Message-level security allows for entire or specific portions of messages to be encrypted using XML Encryption, and using WS-Security to transport such payloads in SOAP.

Additionally, security polices can be configured in XML firewalls to counter these attacks. The XML firewall can monitor the symptoms, and when rejected requests reach a threshold level it can send notifications or deny access to the service. The following set of parameters (these are logical names, and product agnostic) should be configured on the XML firewall to avoid these attacks.

MAX_REQUEST_RATE_FROM_HOST

This parameter represents the maximum number of requests that the firewall should receive per second from a particular IP address.

ATTACK_THRESHOLD_COUNT

This parameter represents the threshold number at which the firewall should activate its exception-management scenarios for a possible dictionary attack.

NOTIFICATION_TYPE

This parameter indicates the type of notification that is required. Different and multiple levels of notification are generally configurable. There could be system notifications and/or e-mail notifications. Most firewalls also provide the ability to integrate with existing operations-management systems like BMC Patrol and so on.

AUTO_SERVICE_DENY

This flag indicates whether the XML firewall should automatically shut down the service if the threshold limit is reached. This flag can be set to true for different attacks.

SERVICE_RESTART_INTERVAL

This parameter is used for automated restart of the service after the specified time interval.

In addition, real-time intrusion-detection systems (IDS) can be integrated with XML firewalls. This allows the audit logs to be fed to the IDS, which have built-in sophisticated algorithms for detecting such patterns of attacks and frauds.

11.6.4.2   Handling Computed Authentication Attacks

The session ID that is generated should be a long key and should have a random-number component to it. If a specific algorithm is used to generate these keys then it should be a nongeneric algorithm. Nongeneric algorithms have a random element, which makes it extremely difficult to regenerate or guess the session keys.

11.6.4.3   Handling Session Hijacking and Fixation Attacks

These are impersonation attacks orchestrated to hijack an authenticated session from a legitimate user. The user may also be given a link with a generated session

ID, and then his session can be used for attacks. Web services are even more vulnerable to these attacks as there is no sophisticated mechanism for session management. The following techniques can be used to address this:

(1) Regenerate the session ID after login. This way the session ID passed on by the attacker is rendered invalid, making difficult to conduct session-fixation attacks.
(2) Check the originating IP address of the login request and any subsequent requests. This ensures that the session is bound to the same IP address used at login. However, this technique fails if the attacker shares the IP address with the user. It will also fail if the IP address of a legitimate user changes during the session.
(3) Use WS Addressing to create a virtual session management by using the MessageID attribute. It is imperative that the MessageID is generated using a sophisticated algorithm. It is also important to create a network-transportable session ID that is mapped internally to the actual session ID. Passing the session ID over the network poses a high risk.
(4) Web service providers use username tokens for authentication, and even SAML tokens are typically used for SSO mechanisms. These tokens should be linked to specific time-outs, which ensure that the user has to re-enter credentials and hence that a hijacked session is only valid for a specific time.
(5) Service providers like payment gateways or institutional banks dealing with corporate payments are going the extra mile by putting in two-factor authentication or smart-card-based authentication.
(6) Bind the session ID to the user's SSL client certificate – a very important and often overlooked issue in highly-critical applications: *each* server-side script must first check whether the proposed session was actually established using the supplied certificate.
(7) Encrypt the data passed between the parties, in particular the session key. This technique is widely relied upon by Web-based banks and other e-commerce services, because it completely prevents sniffing-style attacks. However, it could still be possible to perform some other kind of session hijack.

11.6.5 *Handling Man-in-the-Middle Attacks*

A replay attack is a man-in-the-middle type of attack where a message is intercepted and replayed by an attacker to impersonate the original sender. These are the most difficult DOS attacks to control, as the attacker can flood the service with valid requests. They are also known as spoofing attacks. The use of WS-Security username tokens and WS Addressing for managing routing and message relay also does not directly deal with replay attacks. The Web Service Security Username Token Profile [4] specifically states the following techniques to effectively thwart replay attacks:

(1) It is RECOMMENDED that Web-service producers reject any UsernameToken *not* using *both* nonce *and* creation timestamps.
(2) It is RECOMMENDED that Web-service producers provide a timestamp 'freshness' limitation, and that any UsernameToken with 'stale' timestamps be rejected. As a guideline, a value of five minutes can be used as a minimum to detect, and thus reject, replays.
(3) It is RECOMMENDED that used nonces be cached for a period at least as long as the timestamp freshness limitation period, above, and that UsernameToken with nonces that have already been used (and are thus in the cache) be rejected.

As indicated above, it is imperative to use a timestamp-based approach for any kind of token issue and validation. SAML tokens and WS Addressing Message IDs should also be linked to a timestamp mechanism. This allows applications and the token issuers to revalidate or regenerate the tokens upon expiry. There are various other parameters that can be configured on an XML firewall to prevent replay flooding and DOS attacks. Implementing cryptographic technologies is another solution. Using a public-key–private-key combination would mean that even if an attacker hijacked the messages, they would need the signing key to decrypt them. Additionally, an XML firewall can be used to addresses these attacks by adjusting the parameters below (some of them are repeated from earlier solutions).

REPLAY_MONITOR_FLAG
This flag can be configured on the XML firewall. If it is set to true, the firewall will use a set of elements to check whether the same message is passed through the firewall more than once in a specific time interval.

REPLAY_MONITOR_INTERVAL
This parameter is used to enable the XML firewall to check for potential replayed messages. This time interval should be set to less than the token expiry intervals.

MAX_REQUEST_RATE_FROM_HOST
This parameter represents the maximum number of requests that the firewall should receive per second from a particular IP address. It can be used to limit replay attacks from any particular IP address.

ATTACK_THRESHOLD_COUNT
This parameter represents the threshold number of requests at which the firewall should activate its exception-management scenarios.

NOTIFICATION_TYPE
This parameter indicates the type of notification that is required. Different and multiple levels of notification are generally configurable. There could be system

notifications and/or e-mail notifications. Most firewalls also provide the ability to integrate with existing operations-management systems like BMC Patrol and so on.

AUTO_SERVICE_DENY

This flag indicates whether the XML firewall should automatically shut down the service if the threshold limit has been reached. This flag can be set to true for different attacks.

SERVICE_RESTART_INTERVAL

This parameter is used for automated restart of the service after the specified time interval.

MAX_REQUEST_RATE

This parameter is used to define the maximum number of requests for a service. This is done to ensure that any particular service does not absorb a lot of resources, which would affect the performance of other services.

11.6.6 Handling SOAP Attachment Virus Attacks

The advent of WS Attachments [5] as a standard meant that binary data in various formats could be transported with the underlying SOAP messages. WS Attachments allows the SOAP stack to be used in an almost e-mail-like manner. It can also be used for breaking messages into multiple parts. However, just like e-mail-based virus attacks, an attacker can send spurious attachments with SOAP messages. This can be combated by defining security policies that detail the mechanisms by which attachment processing should happen with Web services. The following parameters can be configured in the XML firewall to implement the security policy.

ATTACHMENTS_PROCESS_FLAG

This parameter indicates whether the XML firewall should allow messages that have attachments. If the flag is set to false, all messages with attachments will be discarded.

ATTACHMENTS_EXCLUSION_FILTER

This parameter lists the types of attachment that need to be barred from entry. It will define the DIME/MIME types for exclusion. This is very similar to e-mail policies wherein attachments like .exe are discarded.

ATTACHMENTS_MAX_SIZE

This parameter defines the maximum attachment size that can be processed by the firewall. This parameter needs to set, as potential attackers could send huge attachments to cause DOS attacks.

11.6.7 Handling Parameter-Tampering Attacks

Parameter tampering is a more sophisticated attack, where the attacker actually targets the integrity of the message, changing specific values like amounts and so on. These attacks can be combated by a variety of techniques, including those below.

11.6.7.1 Signed Messages

The message body or certain parts of the body may be encrypted to preserve the integrity and confidentiality of the message. This can be accomplished through traditional PKI infrastructure or X.509 certificates.

11.6.7.2 Schema Validation

All XML messages that come in through the firewall should be checked for validation and conformance to the underlying schemas. It is recommended that strict validation be enforced to ensure compliance. Additionally, various rules can be set within XML firewalls or even the schemas, which can check for data types and formats, and constrain the data to a defined subset. This validation can be done for both inbound and outbound messages.

The following parameters can be used to enforce schema validation on the XML firewall.

INBOUND_MESSAGE_SCHEMA_VALIDATION
This parameter defines whether the incoming messages have to be validated against the underlying message schema. It is highly recommended that this always be true.

INBOUND_MESSAGE_SCHEMA_VALIDATION_LEVEL
This parameter defines the level of validation that needs to be enforced for schema validation. It is recommended that strict validation of the messages be performed.

INBOUND_MESSAGE_ELEMENT_VALIDATION_RULES
This parameter allows definition of additional validation rules that can be performed on the message element or attributes. This is defined for each element or attribute.

11.6.8 XML Attacks

11.6.8.1 External-Entity Attacks

Successful external-entity attacks could lead to unauthorized access, allow access to files on the host and open up TCP ports. There are various techniques to prevent these attacks:

(1) Use SOAP and XMLRPC implementations that are not vulnerable in their parser configuration to external-entity attacks.
(2) Use SOAP and XMLRPC implementations that do not support external entities.
(3) Most parsers allow the user to explicitly specify the external-entity handler. It is generally recommended to suppress external URI references to protect against malicious data.
(4) Do not expose the user accounts that have access to host commands to external entities.
(5) Create an explicit and protected chain of trusted domains and only accept URLs from these domains.

11.6.8.2   XML Denial-of-Service Attacks

XML DOS attacks are orchestrated by attackers to make the response of services extremely slow or completely unavailable to legitimate users. This is generally accomplished by using various techniques such as flooding and complex requests. These requests look like legitimate ones and force the service to process them.

AUTO_REQUEST_BLOCK
This parameter indicates that all requests from an IP address which is sending spurious messages should be blocked.

ATTACK_THRESHOLD_COUNT
This parameter represents the threshold number of requests at which the firewall should activate its exception management scenarios.

ERROR_THRESHOLD_COUNT
This parameter represents the threshold at which action, which might include notification, is taken when a service starts returning an unusually high number of errors or SOAP faults.

AUTHORIZATION_THRESHOLD_COUNT
This parameter represents the threshold at which action can be taken when an excessive number of HTTP unauthorized/forbidden errors are returned.

CPU_THRESHOLD_LIMIT
This parameter represents the threshold at which action can be taken when message processing takes a large number of CPU cycles.

NOTIFICATION_TYPE

This parameter indicates the type of notification that is required. Different and multiple levels of notification are generally configurable. There could be system notifications and/or e-mail notifications. Most firewalls also provide the ability to integrate with existing operations-management systems like BMC Patrol and so on.

AUTO_SERVICE_DENY

This flag indicates whether the XML firewall should automatically shut down the service if the threshold limit has been reached. It can set to be true for different attacks.

SERVICE_RESTART_INTERVAL

This parameter is used for automated restart of the service after the specified time interval.

MAX_REQUEST_RATE

This parameter is used to define the maximum number of requests for a service. This is done to ensure that any particular service does not absorb a lot of resources, which would affect the performance of other services.

11.6.8.3 Complex-Payload Attacks

Attackers may use complex XML payloads by creating a recursion or deep nesting of XML elements to overwhelm the parser. This should generally be configured by creating a policy around SOAP/XML messages, which should clearly define the following parameters.

MAX_NESTED_LEVELS

This defines the maximum amount of nesting allowed inside a particular element. The XML firewall will check whether the message conforms to this configuration. This parameter can be set as per the schema definitions of the messages in WSDLs.

MAX_ATTRIBUTE_ALLOWED

This defines the maximum number of attributes allowed for a particular element.

MAX_ELEMENT_PER_LEVEL

This defines the maximum number of elements allowed for each level in the tree. This prevents attacks where the attacker simply creates a large XML message without any kind of nesting of elements.

RECURSION_ALLOWED

This parameter allows the policy to define whether recursion is allowed within the XML message. This should be switched on only in specific instances when the underlying schema of the message is extremely complex. It is recommended that defining such complex schemas be avoided.

11.6.8.4   Oversized-Payload Attacks

These attacks are intended to overwhelm the parser by sending a large message. They can be prevented by defining the following parameters in the security policy.

MAX_MESSAGE_THRESHOLD

This parameter sets the maximum size of a message. There can be strict rules governing the parsing of messages greater than the threshold size. The policy can allow specific overrides in exception scenarios.

MAX_REQUEST_PROCESS_TIME

This parameter defines the response time for every request that goes through the firewall. It can be tricky to configure as a lot of this depends on the load and complexity of the underlying service. However, it can be used effectively to monitor requests that take a lot of response time and might be possible candidates for DOS attacks.

11.6.9   Known-Bug Attacks

Enterprises use popular open-source frameworks as part of their solutions. However, open-source frameworks have well-publicized bug lists, which are often exploited by hackers to mount attacks. There are various options to combat these attacks, including:

(1) Continuous monitoring of open-source bug lists, to understand issues and create mitigation plans for preventing attacks.
(2) Specialized support structures for maintenance of open-source frameworks.

11.7   Service Threat Solution Mapping

Table 11.6 provides an overview of the solutions to the attacks and vulnerabilities identified in Chapter 7.

11.8   XML Firewall Configuration-Threat Mapping

Table 11.7 provides the reader with an overview of the parameters and mechanisms that can be used in an XML firewall to counter various threats. These features

Table 11.6 Solutions to service-level threats.

Categories	Attacks	Solutions
Service attacks	SQL injection XPath and XQuery	Sanitizing user input Malicious input validation Pattern-matching filters on XML firewalls Schema and XSL validation of all XML messages Avoiding dynamic query generation Error information sanitization
	Cross-site scripting	Filtering all input parameters Encoding all output based on input Validation for all special characters
	Parameter tampering	Signed messages for implementing message-level security Schema validation for messages Firewall rules for constraints on parameters
	Denial-of-service	Denial-of-service attacks can be effectively countered by configuration rules on XML firewalls Controlled IP and port access CPU monitoring techniques Traffic request monitoring and control
Service commu- nication attacks	Replay	Using a timestamp mechanism in the message design that allows identification of stale messages
	Buffer overflow	Reject any UsernameToken that does not contain nonce and timestamps
	Flood	Using XML firewall configurations to manage replay attacks
	Data integrity	Validating message data against the 'source data' wherever possible
Service end-point attacks	WSDL scanning	Service virtualization allows creation of proxies and masking the underlying URL
	WSDL phishing	WS Addressing can be used to create end-point references. These act as a pointer to the actual resource, without exposing the actual resource Creation of a handshake mechanism for interacting with ad hoc WSDL or Web service to ensure validity

Table 11.6   (*continued*)

Categories	Attacks	Solutions
Service session attacks	Session hijacking	Using random generators and nongeneric algorithms for creation of session keys
		Using WS Addressing for session management
	Session fixation	IP-address checking of requests post-login
		Regenerating session IDs after login
		Using encryption techniques for session keys
		Binding session IDs to the user's SSL client certificate
Service authentication attacks	Forced-entry	Cryptographic technologies like digital certificates, digital signatures, PKI etc.
	Brute-force	Using WS Addressing for session management
	Dictionary	Configuring rules and policies on the XML firewall
	Computed authentication attacks	
Service protocol attacks	SOAP routing	Using WS Addressing for routing messages and designing SOAP intermediaries
	SOAP attachment virus	Filtering attachments with messages with virus scans
		Defining security policies on the XML firewall
Service message attacks	External-entity	Suppressing external URI references to protect against malicious data
	Complex-payload	Using the correct SOAP and XMLRPC implementations
	Oversized-payload	Restricting the size of the XML messages
		Strict schema-validation of messages
		Limiting the number of elements and attributes per message
		Configuring rules and policies on the XML firewall
Service message template attacks	Schema poisoning	Using service virtualization to mask the underlying URLs and protect the schemas
	Coercive schema	Strict schema validations
		Read-only access protection to schemas

Table 11.7   XML firewall configuration-threat mapping.

Categories	Attacks	XML firewall configuration parameters
Service attacks	SQL injection	Regular pattern-matching filters on XML firewalls. These filters look for specific patterns and replace the ' character with the " character
	XPath and XQuery	XSL validators for all XML messages validating formats, lengths and special characters
	Cross-site scripting	Creating constraint-based filters that allow the firewall to validate element values against a specified and limited list
	Parameter-tampering	INBOUND_MESSAGE_SCHEMA_VALIDATION INBOUND_MESSAGE_SCHEMA_VALIDATION_LEVEL INBOUND_MESSAGE_ELEMENT_VALIDATION_RULES
Service attacks	Denial-of-service	MAX_REQUEST_RATE_FROM_HOST ATTACK_THRESHOLD_COUNT NOTIFICATION_TYPE AUTO_SERVICE_DENY SERVICE_RESTART_INTERVAL AUTO_REQUEST_BLOCK ERROR_THRESHOLD_COUNT AUTHORIZATION_THRESHOLD_COUNT CPU_THRESHOLD_LIMIT MAX_REQUEST_RATE
Service communication attacks	Replay	REPLAY_MONITOR_FLAG
	Buffer overflow Flood Data integrity	REPLAY_MONITOR_INTERVAL MAX_REQUEST_RATE_FROM_HOST ATTACK_THRESHOLD_COUNT NOTIFICATION_TYPE
Service end-point attacks	WSDL scanning	Using service virtualization and creating service proxies to mask the WSDL
	WSDL phishing	ALLOWED_REQUEST_IP_ADDRESS_LIST
Service authentication attacks	Forced-entry	Configuring security-realm policies to support various authentication and authorization techniques like WS-Security, SAML and cryptography

Table 11.7 (*continued*)

Categories	Attacks	XML firewall configuration parameters
	Brute-force	Creating rules around policy enforcement to ensure complete adherence to defined policies
	Dictionary	Configuring rules and policies on the XML firewall
	Computed-authentication	MAX_REQUEST_RATE_FROM_HOST
		ATTACK_THRESHOLD_COUNT
		NOTIFICATION_TYPE
		AUTO_SERVICE_DENY
		SERVICE_RESTART_INTERVAL
		AUTO_REQUEST_BLOCK
		ERROR_THRESHOLD_COUNT
Service protocol attacks	SOAP routing	ATTACHMENTS_PROCESS_FLAG
	SOAP attachment virus	ATTACHMENTS_EXCLUSION_FILTER
		ATTACHMENTS_MAX_SIZE
Service message attacks	External-entity	MAX_NESTED_LEVELS
	Complex-payload	MAX_ATTRIBUTE_ALLOWED
	Oversized-payload	MAX_ELEMENT_PER_LEVEL
		RECURSION_ALLOWED
		MAX_MESSAGE_THRESHOLD
		MAX_REQUEST_PROCESS_TIME
Service message template attacks	Schema poisoning	Using service virtualization and creating service proxies to mask the WSDL
	Coercive schema	ALLOWED_REQUEST_IP_ADDRESS_LIST
		INBOUND_MESSAGE_SCHEMA_VALIDATION
		INBOUND_MESSAGE_SCHEMA_VALIDATION_LEVEL
		INBOUND_MESSAGE_ELEMENT_VALIDATION_RULES

are generally available out of the box with XML firewall products. This overview is meant to be representative and the parameters might possibly have different names in the various XML firewall products that are available today. However, the table does provide a guide to the general requirements of the XML firewall and how it can be effectively used to counter the various service-level threats and vulnerabilities.

11.9 Conclusion

In this chapter we have explored the role of standards in addressing key SOA security requirements. Further, we explored in detail the prominent SOA security standards in practice, with a view to helping practitioners separate the practical and popular ones from the rest. We have also explored popular emerging security architectures to address the needs of SOA, including XML- and SOAP/WSDL-level attacks/vulnerabilities and the loosely-coupled requirements of message-level security and policy infrastructure. In particular, we explored logical options for configuring XML firewalls to address the specific requirements for handling each security threat/attack type.

References

[1] Spec of XML Encryption, http://www.w3.org/TR/xmlenc-core/
[2] The Platform for Privacy Preferences 1.0 (P3P1.0) Specification, http://www.w3.org/TR/P3P/
[3] Padmanabhuni, S., Singh, V., Senthil kumar, K.M. and Chatterjee, A. (2006) Preventing Service Oriented Denial of Service (PreSODoS): A Proposed Approach. International Conference on Web Services, ICWS 2006, September, pp. 577–84.
[4] Web Services Security Username Token Profile, http://www.docs.oasis-open.org/wss/2004/01/oasis-200401-wss-username-token-profile-1.0.pdf
[5] WS-I Attachments Profile Specification, http://www.ws-i.org/Profiles/AttachmentsProfile-1.0-2004-08-24.html

Further Reading

CISCO ACE Gateway User Guide, http://www.cisco.com/univercd/cc/td/doc/product/Webscale/ace_xml/xml_5_1/xmlgs.pdf
Erradi, A., Maheshwari, P. and Padmanabhuni, S. (2005) Towards a Policy-Driven Framework for Adaptive Web Services Composition. International Conference on Next Generation Web Services Practices, NWeSP, 2005, August 22–26, 2005.
Extensible Access Control Markup Language 3 (XACML) Version 2.0, http://docs.oasis-open.org/xacml/2.0/access_control-xacml-2.0-core-spec-os.pdf
IBM Datapower User Guide, http://www.redbooks.ibm.com/redpieces/pdfs/redp4365.pdf
IBM Web Services Tool Kit, http://www.alphaworks.ibm.com/tech/Webservicestoolkit
Liberty Specifications for Identity Management, http://www.projectliberty.org/resource_center/specifications
OASIS Web Services Security (WSS), http://www.oasis-open.org/committees/wss/

Project Liberty, http://www.projectliberty.org/

Schmid, A., Padmanabhuni, S. and Schroeder, A. (2007) A Soft Constraints-Based Approach for Reconciliation of Non-Functional Requirements in Web Services-Based Multi-Agent Systems, IEEE International Conference on Web Services (ICWS 2007), pp. 711–18.

Spec of XKMS, http://www.w3.org/TR/xkms/

Web Services Federation Language (WS-Federation), http://schemas.xmlsoap.org/ws/2003/07/secext/

Web Services Policy Framework (WS-Policy), http://specs.xmlsoap.org/ws/2004/09/policy/ws-policy.pdf

Web Services Secure Conversation Language (WS-Secure Conversation), http://specs.xmlsoap.org/ws/2005/02/sc/WS-SecureConversation.pdf

Web Services Trust Language (WS-Trust), http://schemas.xmlsoap.org/ws/2005/02/trust/

WS-Authorization, http://www.w3c.or.kr/~hollobit/roadmap/ws-specs/WS-Authorization.html

WS-Privacy, http://www.serviceoriented.org/ws-privacy.html

Xtradyne Firewall Architecture, http://www.xtradyne.com/products/ws-dbc/WSDBCarchitecture.htm

12

Case Study: Compliance in Financial Services

12.1 Introduction

'Compliance is a journey not a destination.' It requires constant monitoring and vigilance, combined with a balancing of cost, risk and transparency.

In this case study we look at various aspects of the financial industry. Financial frauds have prevailed historically, and been omnipresent in multiple forms. There have been many cases, such as the Barings Bank/Nick Leeson fraud [1] of the late 1980s, or the Allied Irish Bank/Allfirst fraud of the beginning of this millennium. Then there were Enron, Tyco and Worldcom in the 1990s and early 2000. As recently as this year, the French banking giant Societe Generale [2] had to write down losses to the tune of billions of US dollars in associated bank fraud.

These cases have led to specific compliance laws like the Sarbanes-Oxley (SOX) Act [3], Basel II [4] and the Uniting and Strengthening America by Providing Appropriate Tools Required to Intercept and Outburst Terrorism Act (USA Patriot Act) [5] being promulgated to enhance accountability and visibility in the banks. Today, banks have to comply with a multitude of regulatory laws, thereby increasing the effort and time spent in addressing compliance issues, not to mention the associated cost aggravation. Organizing reports pertaining to these regulatory compliances requires data from diverse applications, which operate in silos. Further, these applications do not interoperate; hence the effort to cater to a specific compliance need is both time-consuming and costly. Aggravating this is the lack of possibility of being able to reuse data/information across multiple regulatory needs. Existing solutions do not provide a complete end-to-end automation of regulatory compliance in a cost-effective way.

Compliance is about providing the regulators with the expected information in the prescribed format. Since information presently resides in digital form

Distributed Systems Security   A. Belapurkar, A. Chakrabarti, S. Padmanabhuni, H. Ponnapalli, N. Varadarajan and S. Sundarrajan © 2009 John Wiley & Sons, Ltd

in highly-computerized environments, organizations are forced to provision it through specific IT frameworks. IT organizations are being tasked with establishing mechanisms for more effective, systematic control of fundamental business processes, even when compliance issues cut across national and continental boundaries. Thus, irrespective of business size or industry, compliance is becoming a primary concern for CIOs and CTOs at virtually every organization we work with. An increasing focus on transparency, reporting and risk mitigation indicates that the growing demand for compliance capabilities will not plateau in the near future.

Laws typically mandate monitoring provisions; for example, Section 404 of the SOX Act [3] governs the management's assessment of the internal controls of the processes, system and services that produce the enterprise's financial reports. The people, processes and IT applications responsible for the creation, processing, maintenance and production of financial reports and data come under the scrutiny of the Act. It essentially requires managing the data with the assurance of data-quality attributes including integrity, security, traceability and auditability.

During the building of legacy systems, very little attention was paid to the data-quality issues. These systems were designed and operated as per the needs of the business in an ad hoc manner. Over time, such systems developed inconsistencies, and on many occasions caused replication of the stored data. Adding to this, many new business applications and processes were developed and defined, leading to cluttering of the stored data, with no clear distinction of ownership. In spite of the emerging technologies, the information for reporting principally depends on the integrity and security of the legacy data. In order to establish data quality, the applications typically depend on commercially- and custom-built products, which are unpredictable. Needing to comply with reporting regulations in a limited time, organizations have adopted technologies like data warehouses, data marts and document management systems that only provide reconciliation of data.

The Societe Generale incident [6] reemphasizes the effects of lack of fraud-control security. Here the employee was previously a member of the IT department and hence had an in-depth knowledge of the IT systems and processes around fraud and security. The employee used fake e-mail messages for confirmations to justify missing transactions. There was also a process violation where user credentials were borrowed and shared to perform trading under the guise of other employees.

Compliance projects face immense integration challenges. Despite the increasing attention paid to compliance as a pervasive business concern, technical efforts to address the various challenges posed by compliance requirements are being undermined by a myopic focus on tactical initiatives. The typical IT organization is addressing compliance reactively.

Apart from reporting during audits, there are laws like Section 215 of the Patriot Act that require real-time reporting capabilities. The Act mandates the financial enterprise to know its customers well and identify any suspicious activities, including those indicative of money laundering. It directs firms to have robust and extensive systems for customer data monitoring and reporting. In general it attempts to track and limit the financial resources fueling terrorist activities by imposing requirements on financial institutions, with severe penalties for noncompliance.

Until recently, to establish IT governance, many organizations engaged in developing internal control frameworks. But then, due to the increased necessity of assurance in IT systems, confusion prevailed at the existence of multiple frameworks and multiple evaluation methods. With the passing of the SOX Act in 2002, the need for an integrated approach toward IT management and control was realized. After much discussion and interpretation, the Committee of Sponsoring Organizations of Treadway Commission's (COSO) internal control integrated framework was accepted as the de facto control framework [7]. Since COSO's framework was too generic in nature, subsequently many other frameworks have evolved specifically for use with IT-related controls, such as Control Objectives for Information and Related Technology (COBIT) [8], ISO/IEC 17799 and Information Technology Infrastructure Library (ITIL). Among these, COBIT has become the most widely-accepted IT governance and control framework. A typical COBIT-based IT compliance structure [9] is shown in Figure 12.1.

The key requirement for a COBIT-based approach to compliance is reliance on the right kind of processes under the appropriate domain. From an automation and real-time information-access perspective, it is vital that the processes be automated, and that they interact with information from the underlying applications via standards-based services [9, 10]. It is in this context that we illustrate a key requirement for SOX, in terms of the appropriate architecture and the constituents, to enable realization of the information needs for SOX.

12.2 SOX Compliance

Companies that are focused on remaining productive and competitive understand that customers, partners and employees all need deeper access to their organization, giving them what they need at the right time. Doing this effectively and in real time means managing a multitude of user identities and interacting with a variety of systems in an environment of constant change – all while keeping quality of service high and the enterprise secure.

SOX is a broad Act that addresses a number of issues. The most relevant requirements are the following [11]:

(1) CEOs and CFOs must attest to the accuracy of financial statements and disclosures in the periodic report (Section 302).
(2) Companies are responsible for having adequate internal control structure and procedures for financial reporting. Management must assess these internal controls (Section 404).

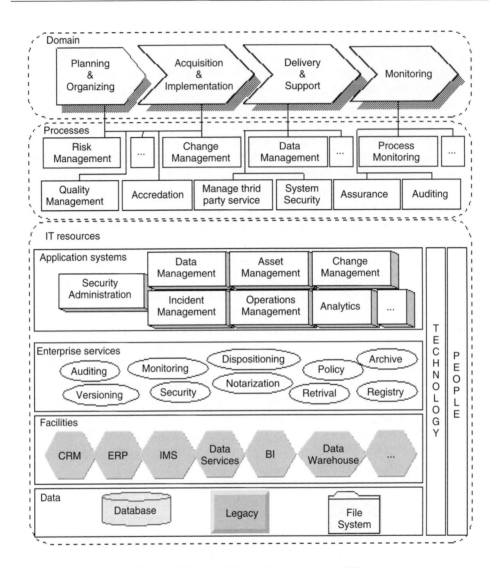

Figure 12.1   An IT compliance structure [1].

(3) Companies must provide real-time disclosures of any events that may affect a firm's stock price or financial performance within a 48-hour period (Section 409).

(4) Companies must protect and retain financial audit records (Section 802).

The SOX security solution should help address key challenges faced by commercial organizations working to comply with multiple sections of the SOX Act. These are illustrated in Table 12.1.

The corresponding issues are dealt with in detail in the following sections.

Table 12.1 Security requirements matrix for SOX compliance.

Section	Requirements	Security implications
Section 302 – corporate responsibility for financial reports	Executives must certify and assume responsibility for the accuracy in the financial reporting	Data-integrity controls
		Data encryption
		Data audits
Section 404 – internal assessment of processes and systems	Executives must engage in audits to check for the effectiveness of internal controls	Access controls
		Enterprise authentication
		Audit controls
Section 409 – real-time issuer disclosures	Near real-time dissemination of any information affecting the financial state of the company to the public	Audit processes and controls
		Fraud detection and control
		Anti-money laundering systems
		System resilience against denial-of-service attacks
Section 802 – protection of financial and audit records	All financial data and audit trails must be securely stored	Secure storage
		Protection against data tampering
		Fraud detection and analysis

12.2.1 Identity Management

The needs are summarized below:

(1) Secure, automated and simple processes to add, suspend and remove users from enterprise systems.
(2) Processes to view, change, audit and report on all user identities, user roles and access privileges.
(3) Processes for audit of federated identities, roles and access privileges across enterprise boundaries.

(4) Replacement of time-consuming, expensive and error-prone processes with a secure one-step process to add or remove users to/from systems.
(5) Workflow capabilities which allow mandatory corporate-approval processes to be enforced and audited.

12.2.2 Policy-Based Access Control

The needs are summarized below:

(1) Access control to systems, information and resources must be strictly managed through policies.
(2) It should be possible for these policies to be enforced, tracked and audited.
(3) Single sign-on must be permeated from the desktop to the applications, hence enforcing a single universal identity policy for access to systems.
(4) Internal and external services and service registries should also be governed by strict access-control rules. This can be accomplished via open standards like SAML.
(5) Access to applications and information should be centrally managed via policy, providing a single point of policy enforcement and audit of access for all users.

12.2.3 Strong Authentication

The needs are summarized below:

(1) A universal, preferably Web-based authentication using various identity mechanisms like usernames/passwords, SAML, digital certificates, smart cards and even biometric devices.
(2) Encryption techniques for all sensitive data, files, folders and e-mail messages.
(3) Only strong authentication mechanisms should be employed, including real-time challenge–response techniques for access to sensitive information.
(4) Authentication mechanisms should exist across various channels, such as PDAs BlackBerrys and so on.

12.2.4 Data Protection and Integrity

Internal controls around data access and data integrity can be enforced through the use of encryption and digital signatures, respectively. Data contained in files, folders or e-mail messages can be encrypted to prevent unauthorized access due to security breaches or weak access controls. That same data can be digitally signed to provide both transaction accountability and data integrity, supplying organizations not only with information on who signed the data, but also with verification that it did not change from the time it was signed, regardless of whether it traveled across the Internet or was stored locally.

12.3 SOX Security Solutions

The compliance solution to the impact of regulations like SOX is generally a multipronged approach which addresses the following facets:

(1) people
(2) processes
(3) technology.

The Societe Generale incident [2] proves that processes need to be in place to prevent internal frauds. Also, there need to be strong governance mechanisms to ensure the execution and enforcement of the processes and policies.

COBIT [8] represents a detailed control framework for IT organizations. The COBIT control definition contains a detailed IT-oriented framework consisting of 4 major domains, 34 IT processes and 318 control objectives. Full compliance with this specification is a difficult task for most IT organizations. However, it serves as a best-of-breed control-environment definition with more detailed guidance and is probably the best starting point for any enterprise.

An important step in a SOX compliance plan is to evaluate the current state of readiness. Constructing a global compliance plan is a critical SOX compliance step. Within a global plan, measured steps are needed to ensure systematic control processes. A global plan must address the feeder and core financial systems and the surrounding enterprise and IT processes that are used to plan, execute and control financial system operation.

The following are recommendations for the people, process and technology dimensions.

12.3.1 People

(1) *governance:*
  (a) stakeholder commitment and buy-in
  (b) creation of a global and regional compliance and security governance council
  (c) identification and creation of accountable roles for compliance managers, who should be responsible for regional and global compliance
  (d) identification of assets (people) which are high-risk in terms of access to sensitive data
  (e) creation of a whistleblower policy with secure and guaranteed anonymity
  (f) application and system ownership to ensure proper accountability of system and information audits
(2) *awareness training:*
  (a) role of IT security in SOX compliance

(b) facets of IT security in SOX

(c) awareness of the compliance program and how it will be executed.

12.3.2 Process

The following are some of the processes that need to be addressed for SOX compliance:

(1) global compliance process
(2) communication of the compliance-process management and execution
(3) planning and strategy processes
(4) governance and execution for security management
(5) whistleblower policy
(6) fraud identification and management
(7) extension of the security process to suppliers and partners
(8) physical and infrastructure security management
(9) business processes security and management
(10) project planning
(11) project tracking and management
(12) vendor and subcontractor management.

12.3.3 Technology

(1) definition of a global security policy, including compliance requirements.
(2) identification and classification of IT systems:
   (a) core transaction systems
   (b) core satellite systems
   (c) partner data and feed systems
   (d) other enterprise systems
(3) mapping of processes to security policies
(4) definition of centralized authentication mechanisms
(5) secure password policies and password management
(6) defininition of secure access control lists
(7) single user identity across all systems, with appropriate access control mechanisms
(8) automated identity provisioning and deprovisioning, allowing easy on-boarding of new employees, contractors and so on
(9) automated workflow-based mechanisms to removes access of subcontractors and employees upon termination
(10) timely and detailed system and process audits
(11) records-management policy defining retention policies, recovery and so on
(12) e-mail security and retention policy
(13) information security and data administration policies.

12.4   Multilevel Policy-Driven Solution Architecture

In the context of SOA-enabled reference architecture for handling the aforementioned technology dimension of the compliance requirements of banks, we have witnessed the roles of different kinds of policy and a lifecycle approach to these policies. These need to be brought together in handling diverse compliance needs.

The following are the components of the policy-driven architecture:

(1) password policy manager
(2) identity provisioning and deprovisioning
(3) centralized policy-based access manager:
    (a) policy-based access control
    (b) role-based access control
    (c) rule-based access control
    (d) fine-grained access control
    (e) entity-based access control
(4) data/information access control
(5) policy-enforcement adaptor
(6) policy repository
(7) policy workbench for defining policies
(8) rules engine
(9) user-identity manager
(10) real-time event and activity monitoring
(11) policy-change audit control and reporting
(12) real-time auditing and reporting
(13) dynamic workflow and process management
(14) support for federated identities
(15) support for business-partner-entity hierarchies to support seamless federation.

There is a gap between the 'documented' security policy and the actual 'executed' run-time policy. Policies can coexist at multiple levels, warranting various compliance requirements (Figure 12.2). Some of these policies are listed below:

(1) business-level policies
(2) application-level policies
(3) information/data-level policies
(4) resource-level policies.

Business-Level Policies

These policies can generally be extremely generic, but attempt to define the business rules that control the behavior of a process or system. It is typically these business rules that require analysis, for operational as well as security

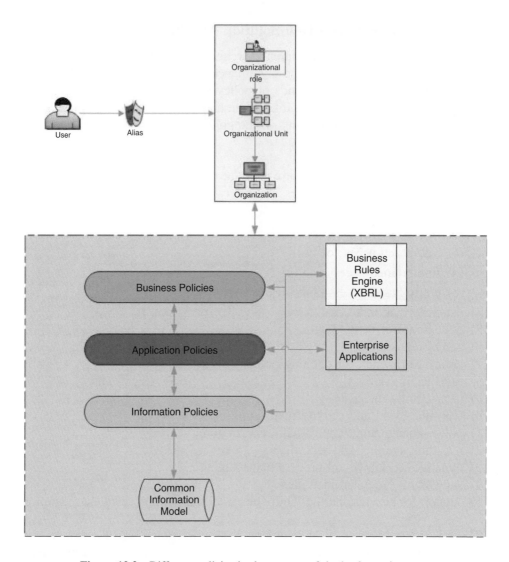

Figure 12.2   Different policies in the context of the bank requirement.

reasons. There are various systems, like fraud monitoring and anti-money laundering, that feed off this data, to monitor and alert us to any potential risks.

Application-Level Policies

These are specific policies, like access control and information security, that need to be defined and implemented to ensure that the right person sees or manages the right data for each application. Hence these policies work with the overall RBAC or access-control lists and identity-management policies.

Information/Data-Level Policies

These policies define the relationships between the bits of information that can be viewed or modified by users or roles. They deal with granular data entities as well as informational entities like reports and so on. They can be defined on top of a Common Information Model enforcing policies at an entity and an attribute level.

We suggest a policy-based middleware approach that can integrate multiple types of policy, as stated above.

12.4.1 Logical Architecture and Middleware

Figure 12.3 shows a policy-driven architecture for managing access control. The solution uses different policies, which are defined at design time, to enforce the

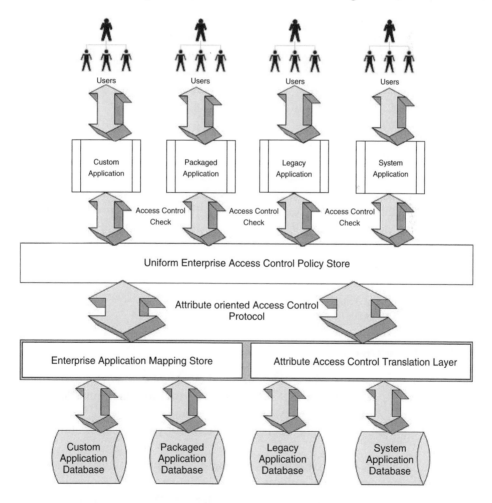

Figure 12.3 Policy management for compliance architecture.

appropriate access control at run time. Extensible access-control mark-up language (XACML) can be used as the policy-definition language. The business policies can be defined using a combination of business rules mark-up language (BRML) and land-extensible business reporting language (XBRL). All the policies are maintained in the policy repository. By utilizing a central point of authentication and role- and rule-based access control, the architecture ensures that security policies can be centrally enforced, resulting in improved security and simplified management. It keeps track of intrusions and unauthorized access activity with real-time audit of any such events. It also creates well-defined, repeatable and auditable security processes that can be enforced enterprise-wide based on user identities.

The applications use a common enterprise security service to request access to any resource. The request types can be different depending upon the policy that needs to be asserted:

- For a coarse-grained resource like access to a specific business service.
- Application of business-rule policy like checking the approval limit for a particular transaction for a user.
- Information access policy like access to a report or even extremely fine-grained access control like specific attributes in a report.

When a client makes a resource request upon a server, the security service is implicitly or explicitly mandated and configured to perform access-control checks. The security service is the policy enforcement point. In order to identify and enforce the appropriate policy, as in Figure 11.9, the policy server will formalize attributes describing the requester and delegate the authorization decision to the policy decision point. Applicable policies are located in a policy store and evaluated at the policy decision point, which then returns the authorization decision. Using this information, the policy enforcement point can deliver the appropriate response to the client. Next, the services of the policy decision point entity are invoked to locate an applicable policy, evaluate it, and return an access decision to the policy enforcement point.

12.4.1.1 Policy Chaining

There are many scenarios where a policy needs to invoke another policy in turn. This is known as policy chaining. There might be multiple hierarchy levels with a tree structure. While each policy is atomic in nature, during chaining the root is called the parent policy. The decision of the child policies are bubbled up to the parent.

This can happen when access to an application service implicitly encompasses business policies and information policies that need to enforced. A classic example in the financial domain is the transaction authorization. While a user could possibly have access to the transaction authorization, there might be a policy defined around the specific approval limits for that user. Hence the business policy is implicitly chained. In addition, the user might not have access to specific customer attributes, such as customer address and phone number, as they are a high-value customer. In such cases the information policy defined is also implicitly chained to the parent.

12.5 Conclusion

In this chapter we have presented a detailed but pointed security case study in the context of financial services. The security needs, in the people, process and technology dimensions, for addressing compliance requirements in the context of the SOX Act have been addressed in detail. To the end, a logical, policy-driven architecture was presented, with multiple levels of policy to handle these requirements. While this case study presents only a small facet of the wide gamut of security requirements in distributed systems, it addresses a real, practical and important pain point, from a security point of view, for financial companies and, most importantly, banks.

References

[1] Barings Debacle, http://www.riskglossary.com/link/barings_debacle.htm
[2] Société Générale Uncovers $3.7bn Fraud by Rogue Trader, news item http://www.guardian.co.uk/business/2008/jan/24/creditcrunch.banking
[3] Sarbanes-Oxley Act, http://www.sarbanes-oxley.com/
[4] Basel II: International Convergence of Capital Measurement and Capital Standards: A Revised Framework, http://www.bis.org/publ/bcbs107.htm
[5] PATRIOT Act, http://www.whitehouse.gov/infocus/patriotact/
[6] Poor IT Security to Blame in Société Générale Fraud, http://www.infoworld.com/article/08/02/20/Poor-IT-security-in-Societe-Generale-fraud_1.html
[7] Committee of Sponsoring Organizations of the Treadway Commission, http://www.coso.org
[8] COBIT Framework, http://www.isaca.org/cobit/
[9] Kulkarni, N., Kumarasamy Mani, S. and Padmanabhuni, S. Reckoning Legislative Compliances with Service Oriented Architecture: A Proposed Approach. Proceedings of the 2005 IEEE International Conference on Services Computing, pp. 16–23.
[10] SOA Meets Compliance: Compliance Oriented Architecture, http://www.redmonk.com/public/COA_Final.pdf
[11] Sarbanes-Oxley (SOX) – Impact on Security in Software, http://www.developer.com/security/article.php/320861

Further Reading

Butler, C.W. and Richardson, G.L. Potential Control Processes for Sarbanes-Oxley Compliance, http://www.isaca.org
SOX – Impact on security in Software, http://www.developer.com/security/article.php/3320861
The Sarbanes-Oxley (SOX) Act and the Impacts of Non-Compliance, http://www.entrust.com/governance/sox.htm

13

Case Study: Grid

In this chapter we will discuss a case study in the financial domain. In the highly-competitive world of financial services, it is imperative that valuations of financial instruments and risk numbers be computed quickly, efficiently and in a cost-effective manner. For example, if a portfolio consists of 50 different financial instruments, a 50×50 risk-correlation matrix involving multiple Monte Carlo simulations will result in a significant investment in processing power. Or when pricing CDOs or valuing swaptions, the time for processing needs to be minimized or else the market opportunity could disappear. Hence computing the need for computing power is absolutely essential.

Different models have been used to achieve the ultimate goal of computing more financial elements in less time. Using higher-end computing resources is one; with mainframes and very high-end clusters, calculations can be done faster and business benefits can be obtained. However, the infrastructure cost is substantial and the benefits obtained from getting better performance are largely offset by the expense of maintaining the system. So the enterprises in the financial domain started looking at more and more of the shelf 2–4 CPU hardwares, connecting them together using a high-speed network like 10 Gb Ethernet or Infiniband to provide performance. Grid computing middleware plays an important role here. If we look at such systems, they are like isolated grid environments. These types of solution are prolific in the financial verticals. However, the environment is still very closely locked with specific applications, and hence security is limited. Organizations are seeing the availability of huge numbers of desktops and low-end computing systems, which are either unutilized or grossly underutilized, and asking whether those systems can be brought into the purview of the grid environment. Once desktops and low-end servers used for day-in, day-out jobs are brought into an enterprise-wide grid system which does work beyond the high-end computation that the financial grid does, several security issues come to the fore. In this case study we look at a financial organization which is moving its high-end computation applications, like the options-pricing application, to

Distributed Systems Security   A. Belapurkar, A. Chakrabarti, S. Padmanabhuni, H. Ponnapalli, N. Varadarajan and S. Sundarrajan © 2009 John Wiley & Sons, Ltd

an enterprise-wide grid. We look at the security issues, and at existing solutions which can help.

It is to be noted that high computation needs are not confined to the financial industry. The energy trading and risk management (ETRM) industry is also in constant need of high computation power, for use in the following areas:

(1) *Complex calculations:* for example, real-time position updates with MTM calculations every time a deal in entered, cancelled or edited; value-at-risk (VaR) calculations (by Monte Carlo simulations); scenario analysis; back testing and risk metrics aggregations; portfolio optimization; and hedge calculations.

(2) *Batch applications:* many organizations have applications which are functionally great for front-office requirements (option calculators, pretrade analytics, etc.) but are simply not used for trading since they take a very long time to give out results due to limitations in computing power; these applications end up being used only in mid-office to get batch results.

(3) *Data heterogeneity:* there are multiple data sources, data types and geographies. ETRM data includes market data (prices, trade quotes, news, instrument data, etc.), static data (some instrument characteristic data), reference data and the firms' own trade data. All these can be available at different geographical locations, and some may have to be calculated for further use. For example, a typical portfolio of an investment bank will have trades executed in different geographies. Many of the data items from across geographies could classify as risk factors and are possible candidates for the variance–covariance (VCV) matrix, also called the covariance matrix calculation. Sometimes, due to the geographical nature of the instruments and also the fact that the portfolios are independent, the VCV matrix for a regional portfolio can be computed at the region itself. So servers can be designated to perform specific calculations. This can achieve two things: (i) a centralized control of such risk calculations; (ii) a deemphasis on the need for centralization of data.

(4) *High-volume data:* high-frequency data from multiple sources has to be correctly routed for best execution, for algorithmic trading, for real-time risk analytics and so on. Since the data is high-frequency, any kind of analysis will be very computationally intensive.

13.1 Background

In this section we will look at a typical example of an enterprise grid computing system. The enterprise we are looking at here is a financial services company with offices in the United States, Europe and India. Before the integration of grid, the IT infrastructure looked like this:

(1) There were different clusters across the globe. The clusters were provisioned at their peak usage and hence grossly underutilized.

(2) The enterprise had made a considerable effort to consolidate user identities through Windows Active Directory. However, users in different clusters were using multiple user accounts.
(3) The workload in the clusters was mainly composed of highly compute-intensive jobs of the credit analysis and long-running batch types. Since most of the batch jobs were running at night, resource sharing across geography was considered a very effective mechanism for increasing resource utilization. (Hence grid was the obvious choice.)
(4) One of the clusters computed the critical options-pricing application, which is very important to the organization as a whole. Integration of this application with other batch applications is critical, as the performance issue is key.

The requirements that the enterprise had in moving toward a grid-based system were as follows:

(1) There should be a centralized identity-management system that tracks the users submitting jobs to different clusters. It should be combined with the centralized monitoring mechanism, which can track usage per department and per group.
(2) The huge pool of enterprise desktops should mainly be used for the batch jobs. Since the jobs will be coming from different departments, the desktops need to have sandboxing mechanisms.
(3) Very elaborate policy mechanisms are needed to bind the users, applications, resources and different metrics.
(4) Since the enterprise had made a significant effort to integrate service-oriented architecture (SOA), all solutions should be compatible with SOA standards.
(5) There are multiple grid/cluster interfaces for submission of different applications. There should be single sign-on (SSO) and an authentication mechanism for all the different interfaces.
(6) Finally, the *performance* of the options-pricing application should not suffer.

13.2   The Financial Application

The application that we are considering for integration with the grid system can be called 'interest option attributes estimation'. It uses two different methods to calculate premiums of financial derivative, called swaption. The first is the Black–Scholes method. The second is the trinomial tree option model, which is an extension of the binomial option pricing model. All possible swaptions are considered and some of their attributes are estimated, such that the premium values calculated by the two methods are as close as possible. While doing this, we try to estimate the parameters of mean reversion and instant volatilities at all the points in the time grid. Finally, we recalculate implied volatilities from the premiums calculated through the latter method, using the Black–Scholes method in reverse. The program inputs a standard yield curve, swaption volatility values and an array of booleans to estimate factors such as mean reversion and instant volatility, so that the premiums are calculated by the tree-based approach for option pricing.

The application takes input in the form of the three files. These are processed to get the two outputs.

- *Inputs:* a yield curve, base volatilities, Cartesian grid.
- *Outputs:* discount factors, mean reversions and instant volatilities at various time points; implied premiums and volatilities of swaptions.

Figure 13.1 shows a high-level overview of the grid architecture. The architecture consists of a scheduler, which distributes the applications on to the grid infrastructure, and the distributed cache, which is used to prefetch the data from the data store and temporarily store it to be used by jobs running on each grid node. The components are described in detail below:

(1) *Parallel applications:* individual tasks can be parallelized using parallel programming paradigms like message passing interface (MPI) or parallel virtual machines (PVMs). Another approach is to run each application on divided data sets, if the applications are data-parallel in nature.

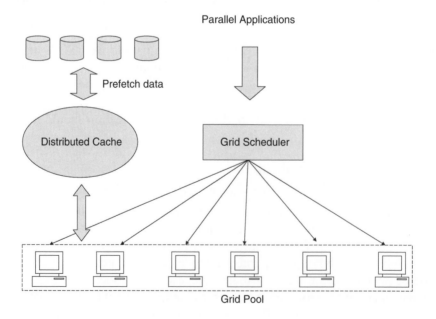

Figure 13.1   Grid architecture.

(2) *Scheduler:* this is responsible for scheduling the job on to a grid node. Different policies based on CPU/memory utilization of machines and resource availability can be used. For scheduling purposes, open-source schedulers like CONDOR or proprietary solutions like Platform LSF can be used.

(3) *Distributed cache:* data can be prefetched from the data sources and placed in the distributed cache. Different distributed cache products like Gemstone Gemfire or Gigaspaces can be used for this purpose. It is to be noted that Gigaspaces is based on Javaspaces specifications and includes workers running on the grid nodes. The advantages of having the distributed cache are manifold. First, the data access time is reduced as the data is prefetched. Second, the data sources can be updated in an asynchronous manner, resulting in better performance. Finally, most of the operations can be performed on the data cache itself, so that significant cost-saving is achieved.

13.3   Security Requirements Analysis

The security requirements that determine the architecture of the security solution are: confidentiality, authentication, single sign-on and delegation, authorization, identity management, secure repository, trust management, monitoring and logging, intrustion detection, data protection and isolation, and denial of service.

13.3.1   Confidentiality Requirement Analysis

Confidentiality is one of the most critical requirements in the grid environment, due to its distributed nature. The problem is even more critical in a financial environment, due to the confidential nature of the information that is being acted upon and transferred. There are many places where confidentiality assumes significant importance:

(1) Movement of data from permanent data sources to transient data sources like distributed shared memory.
(2) Secure scheduling of jobs to distributed grid resources.
(3) Movement of result sets or data back to the submitter.
(4) Confidential execution of jobs and handling of data within a subsystem like the grid nodes.

Solution:
In the case study, two different approaches are possible. The first is using the traditional SSL/TLS channels for confidentiality between the different components, and the second is using WS Security. One of the main concerns with using WS Security in a financial environment is that it is too much of a drain on performance. Therefore, we use SSL/TLS for secure movement of information between different components in a confidential manner. The execution of jobs within a

subsystem has not been made confidential, due to the overhead it will introduce. Since border protections are in place, access to individual resources will be restricted, and hence the need for confidentiality during job execution adds to the overhead.

13.3.2 Authentication Requirement Analysis

Authentication is another crucial component of the security architecture of any grid system. The authentication system should be able to authenticate users logging in to the grid system.

Solution:

As mentioned earlier, one of the requirements of the system was the interoperability of the security mechanisms with all the existing interfaces of the enterprise. The design of the system is based on a centralized authentication system, which validates a user's credentials and sends an authentication token back to the grid entry point. The grid entry point can be any interface which redirects the request to the centralized authentication system. The authentication token is signed by the authentication system and hence can be verified at any point.

13.3.3 Single Sign-On and Delegation Requirement Analysis

SSO is critical in any distributed environment as there are multiple systems where one needs to log in to perform certain operations. For example, a user may submit a job which needs to traverse multiple systems before it finally gets executed. Similarly, delegation is critically important, as jobs executing in a node may need the authority to access databases.

Solution:

The solution provides an SSO and delegation capability, thus reducing the number of times a user must enter their pass phrase when multiple resources are used, which is common in a grid scenario. This is done by creating a proxy. A proxy consists of a new certificate (with a new public key in it) and a new private key. The new certificate contains the owner's identity, modified slightly to indicate that it is a proxy. The new certificate is signed by the owner, rather than a certification authority (CA). The certificate also includes a time notation, after which the proxy should no longer be accepted by others. MyProxy is used in the solution for credential management capabilities.

13.3.4 Authorization Requirement Analysis

Like any resource-sharing systems, grid systems require resource-specific and system-specific authorizations. The authorization systems can be mainly divided into two categories: VO-level systems and resource-level systems. Virtual organization (VO)-level systems have a centralized authorization system to provide credentials for users to access resources. Resource-level authorization systems, on the other hand, allow users to access resources based on the credentials the users themselves present.

Solution:

In the proposed solution, VO-level authorization is enforced through the use of authorization tokens based on standards like SAML. A centralized policy repository is used to create the authorization tokens, which can be present in any database. Resource level authorization is enforced through mapping authorization to local resources in various local systems. The policy decision point (PDP) sends the authentication token to the centralized authorization system. The authorization decision is based on the policy information stored in the policy database. The authorization system verifies the authentication token and then consults the policy database and creates an SAML token based on the policies. The token contains information which binds the resources with the roles and applications that run on them. For example, a policy statement can say that Role R is entitled to run jobs on machines X, Y and Z. Similarly, policy information might say that Application A has x number of licenses, or that Application B is only installed on machines X1, Y1 and Z1. The policy database is updated manually by an administrator as well as through an update service, which is linked to the system directory where the identities of the enterprise users are stored.

13.3.5 Identity Management Requirement Analysis

The directory of users is a critical component in any enterprise. In a grid or cluster environment generally, users may have various different user credentials, which all need to be mapped to the enterprise-wide directory system.

Solution:

In the case study, LDAP is used as the identity-management system. Mapping between identities in LDAP and the enterprise-wide grid and cluster systems is synchronized and updated through regular batch processes.

13.3.6 Secure Repository Requirement Analysis

Credentials are of three main types: identity credentials, authentication credentials and authorization credentials. Identity credentials are used to uniquely identify a particular user. Authentication credentials are mainly to authenticate users. Authorization credentials are used to authorize users. They all need to be stored securely, as attackers could subvert a whole grid system if they gained access to certain credentials.

Solution:

The MyProxy system was developed in the University of Illinois, Urbana Champagne (UIUC) to meet the credential-management requirement of the grid community. MyProxy Toolkit is the grid middleware and is quite popular. Though it is a university project, it has been used in major grids, including NEESgrid, TeraGrid, EU DataGrid and the NASA information power grid. In the case study, MyProxy is used for secure storage of credentials.

13.3.7 Trust Management Requirement Analysis

Managing trust is crucial in a dynamic grid scenario where grid nodes and users join and leave the system. Therefore, there must be a mechanism to understand and manage the trust levels of new systems and nodes joining the grid. The trust lifecycle is mainly composed of three different phases: the trust creation phase, the trust negotiation phase and the trust management phase. The trust creation phase generally occurs before any trusted group is formed, and includes mechanisms to develop trust functions and trust policies. Trust negotiation, on the other hand, is activated when a new untrusted system joins the current distributed system or group. The third phase, or the trust management phase, is responsible for recalculating the trust values based on the transaction information, distribution or exchange of trust-related information, updating and storing the trust information in a centralized or distributed manner.

Solution:

The solution assumes that trust is managed through an enforcer (PEP) which identifies trusted nodes. If a complex trust management trust framework is required, PeerTrust can be considered.

13.3.8 Monitoring and Logging Requirement Analysis

In addition to infrastructure-level monitoring, which is carried out at system, cluster and grid levels, host-, network- and application-level monitoring also assume importance in some cases.

Solution:

Monitoring of resources is essential in grid scenarios, primarily for two reasons. First, different organizations or departments can be charged based on their usage. Second, resource-related information can be logged for auditing or compliance purposes. The different stages of monitoring are: data collection, data processing, data transmission, data storage and data presentation. The different monitoring systems available can be broadly categorized as system-based, cluster-based and grid-based. In Chapter 11, we provided details of different monitoring systems.

(1) *System-level:* the system-level monitoring systems collect and communicate information about standalone systems or networks. The simple network management protocol (SNMP) is an example of a system for managing and monitoring network devices. Examples of open-source and popular system-monitoring tools include Orca, Mon, Aide, Tripwire and so on. The case study uses such monitoring tools.

(2) *Cluster-level:* the cluster-level monitoring systems are generally homogeneous in nature and require deployment across a cluster or a set of clusters for monitoring purposes. Popular examples of cluster-level monitoring systems include Ganglia from the University of Berkeley and Hawkeye from the University of Wisconsin–Madison. Ganglia is used as part of the case study.

(3) *Grid-level:* grid-level monitoring systems are much more flexible than the others and can be deployed on top of them. Many of the grid-level monitoring systems provide standards for interfacing with, querying and displaying information in standard formats. Examples of such monitoring systems include R-GMA, Globus Monitoring and Discovery Systems (MDS), Management of Adaptive Grid Infrastructure (MAGI) and GlueDomains. MAGI is used as part of the case study.

All activities being performed by various components (subject to the underlying component supporting them) within the proposed grid system would be logged securely through encrypted log files. Alternatively, log entries could be stored in database tables set aside for this purpose. Access to these log files can be restricted based on the privilege level of users.

13.3.9 Intrusion Detection Requirement Analysis

Detecting the presence of any intrusion is one of the critical requirements of any system, especially a distributed system of this scale.

Solution:
Any complex IDS brings with it the disadvantage of overhead. To make the system less complex, an open-source IDSs like Snort is used. It is supplemented by custom alarm-triggering routines based on information collected in the system.

13.3.10 Data Protection and Isolation Requirement Analysis

One of the key components of the grid system described is the presence of confidential data in multiple systems. Protecting this data is an important requirement of the system. Similarly, the isolation requirement is crucial; we do not want some other application to affect the working of the jobs submitted to a grid node.

Solution:

Application sandboxing, kernel extension and general sandboxing are options available for providing isolation to independent processes in a host. One of the key components in the design of the grid system is the virtualization at its core; this makes it unnecessary to use any sandboxing or kernel extension techniques. The storage released by a virtual machine upon decommissioning needs to purge any prior state to avoid unauthorized access to private customer/application data. Isolation is provided on multiple levels. First, at the firewalls unauthorized traffic is disallowed from the grid network. Second, VO-level authorization prevents unauthorized users from accessing resources. Third, virtualization can provide isolation at the host level.

13.3.11 Denial of Service Requirement Analysis

Denial of service (DoS) is one of the most important requirements of any security system. Since the solution is distributed in nature, the DoS requirements become more stringent.

Solution:

It is to be noted that DDoS attacks cannot be totally prevented, as this is still a major research issue. However, they can be restricted by applying certain best principles:

(1) Filtering of packets to prevent unauthorized users from sending information which may cause DDoS.
(2) Traffic auditing, so that an abrupt rise in traffic level can be detected. This will be based on applications and loads that are available.
(3) Finally, mirroring and redundancy, to prevent one server getting loaded.

Please note that there may be a performance impact if we put too many anti-DDOS techniques in place. A performance/security analysis needs to be carried out to look at the overheads resulting from DDoS implementations.

XML Firewalls: Filtering packets does not provide enough protection against XDoS attacks. It is necessary to be able to understand the XML documents in order to prevent such attacks. XML-level firewalls, as examined in detail in Chapter 11, can understand such documents, and typically look at received SOAP messages or native XML messages to prevent attack. Several companies, such as Reactivity, have developed such firewalls. Once the target Web service is resolved, the XML firewall can apply a stored security policy based on the target address, originating caller identity, message content and, in some cases, the successful execution of prior policies. Most of the common XDoS attacks, such as entity-expansion attacks, can be filtered by adding specific policies at the XML-firewall level. It is to be noted that this type of filtering has a significant effect on performance, as complex policies need to be applied to the incoming XML messages. Therefore, before applying these techniques, performance/security analysis is required.

13.4 Final Security Architecture

Figure 13.2 shows the high-level architecture of the grid security solution for the enterprise. The main components of the architecture include: the authentication system, the authorization system, the monitoring system and the local access and sandboxing system.

(1) *Authentication system:* as mentioned earlier, one of the requirements of the system was the interoperability of the security mechanisms with all the existing interfaces of the enterprise. The design of the system is based on a centralized authentication system, which validates a user's credentials and sends an authentication token back to the grid entry point. The grid entry point can be any interface which redirects the request to the centralized authentication system. The authentication token is signed by the authentication system and hence can be verified at any point.

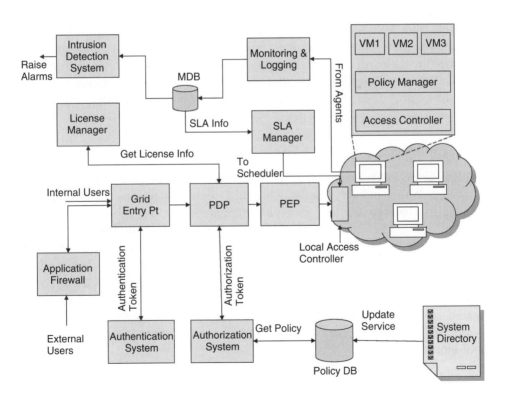

Figure 13.2 High-level architecture.

(2) *Authorization system:* the policy decision point (PDP) sends the authentication token to the centralized authorization system. The authorization decision is based on the policy information stored in the policy database. The authorization system verifies the authentication token and then consults the policy database and creates an SAML token based on the policies. The token contains information which binds the resources with the roles and the applications that run on them. For example, a policy statement can say that Role R is entitled to run jobs on machines X, Y and Z. Similarly, policy information might say that Application A has x number of licenses, or that Application B is only installed on machines X1, Y1 and Z1. The policy database is updated manually by an administrator as well as through an update service, which is linked to the system directory where the identities of the enterprise users are stored.

(3) *Monitoring system:* another important component of the security architecture is the monitoring and logging system. Each node in the grid infrastructure has a monitoring agent which reports any information about it. The monitoring system is very similar to Ganglia (discussed earlier). The information is logged and reported to the reporting interface.

(4) *Local access and sandboxing system:* the centralized policy database stores long-term policy decisions. However, different clusters in different departments may have local policies. The local access controller denies suspended users access to the grid resources. Some other access-control policies can also be implemented in the local system. The local access controller also interacts with the virtual machine scheduler, which submits jobs to the virtual machines within a grid node. Xen is used to create these virtual machines.

13.5   Conclusions

In this chapter, we present a case study of application of grid technology to the financial Services industry. Several financial scenarios (even scenarios in other domains like energy trading) involve computationally intensive tasks like vaR (value at risk) computations, where grid based architectures are the default choice for enterprises. Specific characteristics for these scenarios include the capability of running parallel applications, working with different job schedulers, and distributed caches. In this chapter, we outline the security architecture, by analyzing the detailed security requirements including confidentiality, single sign-on, authorization, identity management and intrusion prevention and detection. We outline a service oriented, grid based loosely coupled architecture to handle these security requirements.

14

Future Directions and Conclusions

14.1 Future Directions

In this chapter we will explore some exciting and promising developments in the context of the future of distributed computing security. We shall concentrate on:

(1) cloud computing security
(2) security appliances
(3) usercentric identity management
(4) identity-based encryption
(5) virtualization in host security.

14.1.1 Cloud Computing Security

Cloud computing, a loosely-defined term, represents a new trend in distributed system development, which involves a combination of multiple tenets. Broadly, it represents a virtualized utility-based model of provisioning computing resources (hardware, software, storage, computing power), provided by extra enterprise providers based on commoditized hardware and software, with the key feature of elasticity (scaling dynamically), and with flexible configuration and vendor choice. Users in cloud computing will typically use multiple service providers, hardware from one provider, storage from another and possibly application services from a third. In a major way, cloud computing realizes the notion of *infrastructure as a service*.

Cloud computing is typically realized via providers making available to end users an abstract standards-based interface, usually a Web service. The set of services available to subscribing end users could include functionalities right from requesting a specific amount of a resource (storage, hardware, etc.) to configuring an image of a resource. Leading vendors, including Amazon, Google, IBM and

Distributed Systems Security   A. Belapurkar, A. Chakrabarti, S. Padmanabhuni, H. Ponnapalli, N. Varadarajan
and S. Sundarrajan © 2009 John Wiley & Sons, Ltd

Microsoft, have all released cloud computing capabilities for end users to make use of their services. The earliest among these was Amazon, which offered storage services (S3), computing capacity in the elastic compute cloud (EC2) services and application services for e-commerce (AWS).

A mainstream opportunity for security product vendors is emerging in cloud computing. Cloud providers can offer a suite of security services, while redirecting all client traffic through shared servers on a subscription model. This is being slowly adopted by some key security vendors. The advantage for cloud consumers is that large security vendors will be able to put large investments into providing a scalable and reliable security service.

Key security concerns in cloud computing include availability, data privacy, compliance and SLA management. In the majority of cloud providers, all the typical SOA issues outlined in Chapter 7 become crucial. However, security in cloud brings additional complexities for a variety of reasons, as outlined below:

(1) Since multiple providers are involved in the cloud, SLA management is complex. In normal systems, SLAs are arrived at between a single provider and the consumer. In a complex cloud transaction with multiple providers, how would SLAs be managed?
(2) Data privacy is another serious concern. How would privacy concerns be addressed by enterprises which wished to store data in the cloud? This could be further complicated by legislative compliance issues.
(3) Since the Web is the platform for delivery of cloud services, they are prone to viruses and other DOS attacks. The lack of large cloud providers is a further cause of concern, as it leads to single points of failure.
(4) The need to work with multiple cloud providers will require seamless and federated approaches to identity management, possibly including usercentric identity management.
(5) Security provisioning is complex in this context. The ability to dynamically provision and de-provision security information is crucial for cloud providers, as enterprise consumers will have a constantly changing user base.
(6) Perimeter security, as it is conventionally applied, loses all meaning. Multiple providers provide a virtual space for sharing data and resources, wherein the concept of a virtual perimeter may need to be re-envisaged.

14.1.2 Security Appliances

Key security functions like encryption and key management are typically computationally expensive. To address the computational complexity of security operations, some vendors offer the possibility of offloading intensive operations on to hardware/firmware, or on to the network layer. Offloading computationally-intensive operations to the hardware layer reduces the overall cycle time in operations involving security. Products which are based on these mechanisms are typically termed as security appliances.

Such operations are more important in the context of higher layers in distributed computing, especially at the service layer and application layer. Most important types of security appliance include secure socket layer (SSL) accelerators, unified threat management (UTM) appliances, XML accelerators, XML networks including application-oriented networks (AON) and XML firewalls (see Chapter 11).

UTM appliances are application-layer firewalls which combine multiple features to handle multiple threats, including e-mail filtering, spam filtering, antivirus capabilities, content filtering and intrusion detection. Some of the popular UTM appliances, including Watchguard [1], Cyberoam [2] and so on, are gaining popularity due to the promise of their addressing multiple Internet threats at the same time. Given the increasing penetration of threats like spam, spyware and viruses, this category of appliance is of prime importance for today's enterprises. Some UTM vendors offer a unique model for helping small and medium enterprises, in the form of a managed service offering, wherein a small appliance is located in the premises but the monitoring and management of threats is done remotely via a shared service center, hence offering security for multiple clients.

SSL accelerators enhance the performance of encryption and decryption tasks on a device, via a coprocessor embedded within them. The round-trip SSL interaction time is greatly hastened, owing to decreased key encryption and decryption time. Typically, the SSL handshake, involving the asymmetric key decryption, is computationally expensive, and is offloaded to the SSL accelerator. An SSL accelerators usually works as a card plugged into a slot on the machine with the Web server, thereby relieving load on the Web-server processors. However, in certain deployments these appliances can be used at a network level by attaching to a network switch, which helps offload the load of SSL processing for multiple Web servers.

XML networks and appliances are typically based on the idea that traditional network-level devices are restricted to introspecting packet-level data, which is insufficient for handling the security requirements at application and higher levels. Application-level security involves introspection of the application interaction payloads, and wire-level data needs to be interpreted at a higher level (application or service level). In view of this, the operations at these higher levels (e.g. WS Security standard implementation, SAML assertions verification) are carried out at a lower network level or on the hardware. These appliances introspect application- or service-level XML content before forwarding it to the next end-point. Additionally, these appliances are typically capable of working with higher-level standards like WS Security, SAML and so on. Finally, the typical XML functions, like transformation, routing and XSL validation, are also carried out. Leading vendors like Cisco's AON [3], IBM's Datapower [4], Intel's Sarvega (now reincarnated as SOA Expressway [5]) and so on have already established products in this area.

In the coming future, with the emergence of Web-scale computing, the computational needs will be larger and newer kinds of appliance will slowly emerge to handle the additional workloads. For example, a specific trend emerging in the enterprise segment is the need to manage employee Internet accesses, leading to employee Internet management appliances [6, 7]. Likewise, a newer category of appliances for handling typical compliance tasks like entitlements control is emerging, as in Rohati [8].

14.1.3 Usercentric Identity Management

Identity management in today's distributed systems is plagued with complex issues like multiple systems of record, multiple authentication systems and complex identity use cases. Existing mechanisms are typically based on username/password combinations; however, we are aware of the huge number of problems associated with password reuse, insecurity of passwords, poor password-management practices, counterfeit sites and so on that open a world of attacks, such as man-in-the-middle attacks and password-theft attacks. Federated identity and single sign-on systems address these concerns partly with better user experience. However, key problems in such systems are the problem of data privacy and the potential of identity fraud due to single point of failure.

To address this concern, a new form of identity management solutions is appearing on the horizon, namely usercentric identity management solutions. Usercentric identity management is predicated upon the notion of the user being in control of their identity. Users can choose the nature and extent of the identity information they wish to present in response to authentication or property requests.

Multiple initiatives have been proposed in the direction of usercentric identity management, including OpenID [9] and Cardspace [10].

OpenID [9] is an Internet-scale emerging de facto standard for usercentric identity. It provides for a decentralized open and free standard for users to control their identity information and use of that information. OpenID works on the idea of decentralized OpenID identity providers (IDPs), and any site registered with any OpenID IDP can log on to the Web site. The popularity of OpenID stems from its growing adoption by leading Internet companies, such as Yahoo!, Google and so on.

Any site relying on OpenID displays the OpenID logo. When a user wishes to log in, it does not present the typical two blanks (for username and password), but instead prompts for one field, namely the user's OpenID identifier. Once the user enters this identifier, which must have been obtained from an OpenID provider, the relying party discovers the identity-provider details from the typed-in identifier, and communicates with that identity provider. The response may vary depending upon the situation: in some cases it will allow for direct interaction without any user intervention; in other cases it will prompt for a further credential from the user, like a password or an information card, following which a request is made to the user to confirm whether the relying party can be trusted. If yes, the control returns to the relying party, along with the user credentials. At this stage the relying site can verify the authenticity of the request or prompt for a further authentication. Once verified, the user is logged in and may carry out transactions.

Likewise, the Windows CardSpace [10] software enables people to maintain a set of personal digital identities, which are shown to them as visual 'information cards'. These cards are representations of their identity that they can use online. The key idea here is to avoid the problems with usernames/passwords and log on to multiple Web sites with one information card. The card can be provided by multiple IDPs, and may be used by multiple relying parties, which all accept it as identity for the user. The key is that the user is in control of the identity interactions and can choose which identity to use with which relying party.

14.1.4 Identity-Based Encryption (IBE)

One of the most interesting advancements in the field of cryptography came from Dan Boneh and Matthew Franklin, of Stanford. Their work was published in 2001 [11] and is called identity-based encryption (IBE). IBE is the solution to the problem floated by Shamir in 1984 [12]: based on a set of global system parameters and a fixed master key, generate a set of private keys corresponding to any set of public keys. The public keys would be used to encrypt messages, which can only be decrypted by the corresponding private key. For example, Alice wants to send a message to Bob using Bob's email address bob@nobody.com. Let G be the set of global parameters known to everybody including Alice, M be the secret master key and P be the private key corresponding to the string bob@nobody.com. It is to be noted that different private keys can be generated corresponding to different strings. Once the private key is generated the message can be encrypted by the string and decrypted using P. This problem was solved by Boneh and Franklin based on the bilinear maps between groups. They showed that Weil pairing on elliptic curves is an example of such a map and implemented the system based on the pairing. The authors proved that the system thus implemented has very strong security properties. They also floated a company called Voltage Systems which implements an IBE-based solution.

At this juncture, the readers may question the usefulness of such a scheme. One of its main advantages is that it frees the sender from having to obtain the public key of the receiver. With IBE, the sender can encrypt their message with any string that can be associated with the receiver, for example an e-mail address or IP address. Hence there is no need to obtain or store the receiver's public key. This becomes important in a bandwidth-constrained environment, where getting public keys through certificates can result in a lot of bandwidth wastage. The authors presented several situations where such a scheme may be very useful:

(1) *Public key revocation:* public key certificates contain a preset expiration date. In an IBE system, key expiration can be carried out by having Alice encrypt a message sent to Bob using the public key: "bob@nobody.com‖current-year". In this way, Bob can use his private key during the current year only. Once a year, Bob needs to obtain a new private key from the trusted private key generator.

(2) *User-credential management:* an IBE system can also be used to generate user credentials. Alice encrypts the message with the following string: bob@nobody.com‖current-year‖clearance. Bob will only be able to decrypt the message if he has the correct clearance; that is, the private key for that string.

The IBE system described above can be very useful for a distributed infrastructure in a constrained environment where security is required. An example of such a system is a sensor grid. Using an IBE system, managing credentials and public keys would be possible in a much less expensive manner than would normally be necessary. However, it is to be noted that such a system is still in research and is not being deployed yet.

Another critical point about an IBE infrastructure is storage of the master key, as the security of the whole system hinges on that. If the master key is compromised, all the private keys need to be regenerated.

14.1.5 *Virtualization in Host Security*

14.1.5.1 **Separation of Management Functions**

With a more widespread use of virtualization technology, the management functions can move to a separate domain, which is completely isolated from the application domain. Segmentation at the network layer through VLANs can provide network isolation as well. Together, they can bring about a natural separation between the management plane and the application plane. Even when one or more applications are compromised, the management components can continue to service and manage other application instances in the application plane.

14.1.5.2 **Protection for Security Agents**

Typically, host-based intrusion detections are strong at identifying any intrusion and are difficult to evade; however, they can be easy to attack, as the operating system is not adequate protection for them. On the other hand, network-based intrusion detections are difficult to attack, but are not as effective as host-based systems at identifying intrusions. Recent research has shown that intrusion detection can be built into a virtual machine monitor on a fully-virtualized platform to give the effectiveness of a host-based intrusion detection system, while retaining the robustness of a network-based one. LiveWire [13] is a reference implementation of such a solution. In future, such capabilities will be bundled and shipped with VMM software.

14.1.5.3   Desktop Security

Virtualization perhaps has the biggest role to play when it comes to desktop security. While the virtualization technology is rapidly advancing through hardware and software innovation in the commodity space, much of this is focused on servers. As the technology becomes friendlier for desktops, there is much to be gained in terms of desktop security. Potentially there could be separate domains for running trusted and untrusted applications, preventing a lot of the host-security and privacy issues discussed in Chapter 4. Besides, some of the security policies (firewalls, isolation levels and device access) can be centrally enforced at the physical-machine level, leaving the guest virtual machine completely within the control of the user, without any security implications.

14.2   Conclusions

In this book, we have taken a holistic look at the security concerns and challenges, processes and solutions, as well as future directions, in distributed systems.

Distributed systems form the backbone of the IT infrastructure in most enterprises today, with mainstream penetration of recent technologies like Web services, grid computing and service-oriented architectures. Security is a key concern in distributed systems and must be approached with a judicious blend of theory and pragmatism in the context of building enterprise-scale IT infrastructures.

We have outlined and described precisely an approach to security engineering, with an exhaustive coverage of the underpinnings of security technologies and the existing and upcoming security threats and vulnerabilities across different layers of modern-day distributed systems architecture. We have detailed this across the four important layers, from bottom to top: *host layer*, *infrastructure layer*, *application layer* and *services layer*. We have specifically avoided the treatment of Web-based security as a separate section, by treating it as part of application security. We have looked at how existing security solutions can be leveraged or enhanced to proactively meet the dynamic and evolving needs of security for next-generation distributed systems.

In the context of host-level security, we have outlined the key threats in the form of transient code threats and resident code threats. The transient code threats can manifest as malware, including worms, spyware, Trojan horses and viruses. Additionally, eavesdropping with malicious intent, for example in collaborative computing such as grids or clusters, can be a key threat at the host level. Likewise, job faults in a grid caused by faultily written programming applications, can bring down hosts and other applications executing on the grid. Similarly, malafide scripts may be injected, causing host shutdown or reboot. Even in cases of genuine use of host computers by a grid application, it is possible that due to excessive use of the CPU by the application, other applications on the host will be starved of compute power.

In addition to the transient code threats at the host level, resident code – code which has been installed with the knowledge of the user – also has the potential to cause host-level attacks, due to the vulnerabilities in it. Typical among these include overflow attacks like the stack-buffer-overflow attacks and heap overflow attacks. Likewise, privilege-escalation attacks are also carried out by unauthorized users who are able to elevate their privilege level and carry out tasks for which they do not have authorization. Sometimes application vulnerabilities leave scope for malicious users to input improper data, leading to injection attacks.

To address the aforementioned host-level threats and vulnerabilities, we have discussed sandboxing and virtualization, two key techniques employed in distributed systems. In sandboxing, access to resources is restricted via the operating system, so that the resident or transient code is not able to do any damage to the host. Virtualization, on the other hand, handles these threats via partitions, each of which is isolated from the others. In today's distributed environments, virtualization is fast becoming a popular technology. It has far-reaching implications for grid and data centers. Existing security products like intrusion-detection/prevention systems will be able to work at each virtual-machine level in a virtualized environment. In addition to sandboxing and virtualization, other techniques like proof-carrying code have been discussed. Proof-carrying code is based on the notion that code needs to be able to prove itself trustworthy, relieving hosts of the headache of preventing code-based attacks. Other techniques for preventing code-injection attacks, such as memory firewalls, have been discussed. These firewalls look at control sections of code and validate the controls against predefined security policies. For viruses and malware, newer categories of product in the form of antivirus and antispyware have been discussed.

Moving up the ladder, in the context of infrastructure-level threats, the usual suspects were discussed first, including DoS, routing attacks, high-speed network threats, wireless network threats and DNS threats. Later, the grid and cluster specific threats were discussed, with focus on architecture-, infrastructure- and management-related issues. In the context of architecture, we focused on the information-security, authorization and service issues. While several network-level threats apply in the context of grid infrastructure too, additional issues arise here in the form of grid network issues, such as integration of grid with existing firewalls. Likewise, in the form of management issues, credential management and trust management were discussed extensively. Storage-level threats were discussed in the context of storage area networks (SANs) and distributed file systems.

To address the aforementioned network-level threats, three primary solutions based on protocols were discussed, including SSL/transport layer security (TLS), virtual private networks (VPNs) and IP Security (IPSec). For addressing DoS and other attacks, techniques like packet filtering and application filtering were discussed. Similarly, for routing attacks, special solutions were discussed using digital signatures. For wireless attacks, solutions like WEP were discussed.

At the grid tier, the security offered by the GSI stack was illustrated. Additionally, specialized solutions, many of which are at the research stage, were proposed in the area of grid services security. Grid authorization systems were discussed at both VO and resource level. For grid infrastructure-level solutions, virtualization-based approaches are showing promise. Likewise, firewall and VPN technology integration is an important area in the context of grid networks. Trust-management systems and credential-management systems were discussed to address management-related threats and vulnerabilities.

For storage-level security, the Fiber Channel Security Protocol (FC-SP) was discussed at length. Additionally, for distributed file systems, existing NFS security models and specialized distributed file systems were discussed.

Moving higher up the tiers, in the context of application- and Web-layer security, primary threats in the form of injection attacks, cross-site-scripting attacks, session-management-related attacks, improper error handling, improper handling of cryptography and insecure configuration issues were discussed. Further, application-layer DoS attacks, overflow attacks and canonical data representation-related issues were discussed.

To address application-level security issues, solutions were proposed in the form of input-validation techniques, centralized validation routines, secure session management, right usage of cryptography, user inputs filtering for cross-site scripting and robust error-handling practices.

At the service level, we outlined key security issues as applicable. In the context of threats, we discussed various categories, including those caused by use of XML as the lingua franca, such as XDOS-, XPath- and XQuery-based attacks. Additionally, the attacks caused by adoption of service-level standards such as SOAP for messaging and WSDL for interfaces were discussed. At the SOAP level, man-in-the-middle and SOAP-virus attacks were discussed. At WSDL level, WSDL scanning and phishing attacks were discussed. Some UDDI-level attacks were also discussed.

To address the abovementioned attacks and to handle service-level security issues, since standards are key and central to SOA, a detailed analysis of the diverse SOA security standards was carried out. Further, a detailed analysis of the adoption levels of different standards was provided. A key emerging infrastructure in the form of XML firewalls was dissected in depth, with a view to exploring how it can address the majority of service-level threats and attacks. We explored some typical deployment architectures of service-level security solutions. In particular, we explored how each of the categories of service-level attack can be addressed by effective use of XML firewalls or similar techniques.

We finished with two case studies exploring the application of the technologies and solutions discussed in the book in real distributed systems that the authors were involved in architecting, developing and running to meet very stringent security requirements. The case studies brought out the practical aspects of designing and architecting security solutions and mapped the described solutions to the distributed systems we typically encounter in workplace.

The first case study dealt with a typical issue in any global banking company, namely compliance. The related issues were explored in detail. We outlined a practical approach to handling the different security requirements for such a scenario, and prescribed a service-based, policy-centered security architecture.

In the second case study, we explored a high-performance grid-based financial services company, which needs to carry out complex calculations in a global grid. The diverse security requirements and issues related to such needs were elucidated in depth. In the end, a working logical architecture addressing the security requirements was explained.

Finally, at the end of the book, we explored some key upcoming trends which will have a significant impact on distributed system security.

References

[1] Watchguard UTM Appliance, http://www.watchguard.com/products/utm.asp.
[2] Cyberoam UTM Appliance, http://www.cyberoam.com/.
[3] CISCO Application Networking Services, http://www.cisco.com/en/US/products/hw/contnetw/index.html.
[4] Websphere Datapower SOA Appliances, http://www-306.ibm.com/software/integration/datapower/.
[5] Intel SOA Expressway, http://www.intel.com/cd/software/products/asmo-na/eng/373233.htm.
[6] Websense – Employee Internet management, http://www.cisilion.com/websense.htm.
[7] Facetime to Offer Granular Control for MySpace, http://www.facetime.com/pr/pr080618.aspx.
[8] Rohati, a high performance entitlement control solution, http://www.rohati.com.
[9] OpenID, http://www.openid.net.
[10] CardSpace, http://www.microsoft.com/net/WindowsCardSpace.aspx.
[11] Boneh, D. and Franklin, M. (2003) Identity-based encryption from the weil pairing. *SIAM Journal on Computing*, **32** (3), 586–615.
[12] Shamir, A. (1984) Identity-based cryptosystems and signature schemes, *Advances in Cryptology, Crypto '84, Lecture Notes in Computer Science*, Vol. **196**, Springer-Verlag, pp. 47–53.
[13] LiveWire, details at http://virtualmachine.searchvmware.com/document;100808/vm-research.htm.

Further Reading

Amazon S3 Service, http://www.aws.amazon.com/s3.
Amazon EC2 Elastic Compute Service, http://www.aws.amazon.com/ec2.
Garfinkel, T. and Rosenblum, M. (2003) A Virtual Machine Introspection Based Architecture for Intrusion Detection. Proceedings of the Network and Distributed Systems Security Symposium, February.

Garfinkel, T. and Warfield, A. (2007) What virtualization can do for Security? *The USENIX Magazine*, **32** (6).

Google Cloud, as exemplified by Google Docs, http://www.documents.google.com.

IBM cloud computing based on Deep Blue technology.

Microsoft Cloud Live, http://www.home.live.com/.

Trend Micro to offer online Internet security service based on cloud computing, http://www.us.trendmicro.com/us/products/enterprise/web-protection-add-on/.

Index

Distributed Systems Security   A. Belapurkar, A. Chakrabarti, S. Padmanabhuni, H. Ponnapalli, N. Varadarajan and S. Sundarrajan © 2009 John Wiley & Sons, Ltd